NATIONAL ACADEMIES *Sciences*
Engineering
Medicine

NATIONAL
ACADEMIES
PRESS
Washington, DC

Innovation, Global Value Chains, and Globalization Measurement

Constance F. Citro, Gail E. Cohen, and Sean H. Strunk, *Rapporteurs*

Innovation Policy Forum

Board on Science, Technology, and Economic Policy

Policy and Global Affairs

Committee on National Statistics

Division of Behavioral and Social Sciences and Education

NATIONAL ACADEMIES PRESS 500 Fifth Street, NW, Washington, DC 20001

This activity was supported by contracts between the National Academy of Sciences and the National Science Foundation (NCSE-1923018). Any opinions, findings, and conclusions, or recommendations expressed in this publication do not necessarily reflect the views of any organization or agency that provided support for the project.

International Standard Book Number-13: 978-0-309-27795-2
International Standard Book Number-10: 0-309-27795-7
Digital Object Identifier: https://doi.org/10.17226/26477

This publication is available from the National Academies Press, 500 Fifth Street, NW, Keck 360, Washington, DC 20001; (800) 624-6242 or (202) 334-3313; http://www.nap.edu.

Copyright 2022 by the National Academy of Sciences. National Academies of Sciences, Engineering, and Medicine and National Academies Press and the graphical logos for each are all trademarks of the National Academy of Sciences. All rights reserved.

Printed in the United States of America.

Suggested citation: National Academies of Sciences, Engineering, and Medicine. 2022. *Innovation, Global Value Chains, and Globalization Measurement: Proceedings of a Workshop*. Washington, DC: The National Academies Press. https://doi.org/10.17226/26477.

The **National Academy of Sciences** was established in 1863 by an Act of Congress, signed by President Lincoln, as a private, nongovernmental institution to advise the nation on issues related to science and technology. Members are elected by their peers for outstanding contributions to research. Dr. Marcia McNutt is president.

The **National Academy of Engineering** was established in 1964 under the charter of the National Academy of Sciences to bring the practices of engineering to advising the nation. Members are elected by their peers for extraordinary contributions to engineering. Dr. John L. Anderson is president.

The **National Academy of Medicine** (formerly the Institute of Medicine) was established in 1970 under the charter of the National Academy of Sciences to advise the nation on medical and health issues. Members are elected by their peers for distinguished contributions to medicine and health. Dr. Victor J. Dzau is president.

The three Academies work together as the **National Academies of Sciences, Engineering, and Medicine** to provide independent, objective analysis and advice to the nation and conduct other activities to solve complex problems and inform public policy decisions. The National Academies also encourage education and research, recognize outstanding contributions to knowledge, and increase public understanding in matters of science, engineering, and medicine.

Learn more about the National Academies of Sciences, Engineering, and Medicine at **www.nationalacademies.org**.

Consensus Study Reports published by the National Academies of Sciences, Engineering, and Medicine document the evidence-based consensus on the study's statement of task by an authoring committee of experts. Reports typically include findings, conclusions, and recommendations based on information gathered by the committee and the committee's deliberations. Each report has been subjected to a rigorous and independent peer-review process and it represents the position of the National Academies on the statement of task.

Proceedings published by the National Academies of Sciences, Engineering, and Medicine chronicle the presentations and discussions at a workshop, symposium, or other event convened by the National Academies. The statements and opinions contained in proceedings are those of the participants and are not endorsed by other participants, the planning committee, or the National Academies.

Rapid Expert Consultations published by the National Academies of Sciences, Engineering, and Medicine are authored by subject-matter experts on narrowly focused topics that can be supported by a body of evidence. The discussions contained in rapid expert consultations are considered those of the authors and do not contain policy recommendations. Rapid expert consultations are reviewed by the institution before release.

For information about other products and activities of the National Academies, please visit www.nationalacademies.org/about/whatwedo.

PLANNING COMMITTEE FOR THE WORKSHOP ON INNOVATION, GLOBAL VALUE CHAINS, AND GLOBALIZATION MEASUREMENT
(Affiliations as of May 2021)

SUSAN HELPER (*Cochair*), Case Western Reserve University
WOLFGANG KELLER (*Cochair*), University of Colorado Boulder
NADIM AHMAD, OECD Centre for Entrepreneurship, SMEs, Regions and Cities
DAVIN CHOR, Dartmouth College
EDUARDO MORALES, Princeton University
ANDREAS MOXNES, University of Oslo
JUSTIN R. PIERCE, Federal Reserve Board
KELLY SIMS GALLAGHER, Tufts University
SARAHELEN THOMPSON, Bureau of Economic Analysis (retired)

PROJECT STAFF
GAIL COHEN, Senior Director, Board on Science, Technology, and Economic Policy (STEP)
SOPHIE BILLINGE, Senior Project Assistant, STEP Board (through June 2022)
CONSTANCE F. CITRO, Senior Scholar, Committee on National Statistics
DAVID DIERKSHEIDE, Program Officer, STEP Board
SEAN STRUNK, Consultant

INNOVATION POLICY FORUM

THOMAS GUEVARA (*Cochair*), Indiana University
DAVID M. HART (*Cochair*), George Mason University
WILLIAM B. BONVILLIAN, Massachusetts Institute of Technology
E. WILLIAM COLGLAZIER, American Association for the Advancement of Science
FRANK DIGIAMMARINO, Booz Allen Hamilton
MARYANN P. FELDMAN, Arizona State University
ANDRE GUDGER, Eccalon
KATHLEEN N. KINGSCOTT, IBM Research
MICHAEL F. MOLNAR (*Ex Officio Member*), National Institute of Standards and Technology
MICHAEL RUSSO, National Institute for Innovation and Technology

BOARD ON SCIENCE, TECHNOLOGY, AND ECONOMIC POLICY

ADAM B. JAFFE (*Chair*), Brandeis University
NOËL BAKHTIAN, Lawrence Berkeley National Laboratory
JEFF BINGAMAN, Former U.S. Senator, New Mexico
BRENDA J. DIETRICH (NAE), Cornell University
BRIAN G. HUGHES, HBN Shoe, LLC
PAULA E. STEPHAN, Georgia State University
SCOTT STERN, Massachusetts Institute of Technology
JOHN C. WALL (NAE), Cummins, Inc. (Retired)
JOHN L. ANDERSON (NAE) (*Ex Officio Member*), National Academy of Engineering
VICTOR J. DZAU (NAM) (*Ex Officio Member*), National Academy of Medicine
MARCIA MCNUTT (NAS/NAE) (*Ex Officio Member*), National Academy of Sciences

STAFF
GAIL COHEN, Director
SOPHIE BILLINGE, Senior Project Assistant (through June 2022)
DAVID DIERKSHEIDE, Program Officer
CLARA SAVAGE, Financial Officer

COMMITTEE ON NATIONAL STATISTICS

ROBERT M. GROVES (*Chair*), Office of the Provost, Georgetown University
LAWRENCE D. BOBO, Department of Sociology, Harvard University
ANNE C. CASE, School of Public and International Affairs, Princeton University, *Emeritus*
MICK P. COUPER, Institute for Social Research, University of Michigan
DIANA FARRELL, JPMorgan Chase Institute, Washington, DC
ROBERT GOERGE, Chapin Hall at the University of Chicago
ERICA L. GROSHEN, School of Industrial and Labor Relations, Cornell University
HILARY HOYNES, Goldman School of Public Policy, University of California-Berkeley
DANIEL KIFER, Department of Computer Science and Engineering, The Pennsylvania State University
SHARON LOHR, School of Mathematical and Statistical Sciences, Arizona State University, *Emerita*
JEROME P. REITER, Department of Statistical Science, Duke University
JUDITH A. SELTZER, Department of Sociology, University of California-Los Angeles,
C. MATTHEW SNIPP, School of the Humanities and Sciences, Stanford University
ELIZABETH A. STUART, Department of Mental Health, Johns Hopkins Bloomberg School of Public Health

BRIAN HARRIS-KOJETIN, Director
MELISSA CHIU, Deputy Director
CONSTANCE F. CITRO, Senior Scholar

Acknowledgments

A planning committee operating under the auspices of the Innovation Policy Forum of the Board on Science, Technology, and Economic Policy, in collaboration with the Committee on National Statistics, convened the workshop, "Innovation, Global Value Chains, and Globalization Measurement" in May 2021. These proceedings are the main product of the workshop. They were prepared by rapporteurs whose charter was to distill the gist of the presentations and the essence of the discussions. The planning committee's role was limited to planning and convening the workshop. The views contained in the proceedings are those of individual workshop participants and do not necessarily represent the views of all workshop participants, the planning committee, or the National Academies of Sciences, Engineering, and Medicine.

This Proceedings of a Workshop was reviewed in draft form by individuals chosen for their diverse perspectives and technical expertise. The purpose of this independent review is to provide candid and critical comments that will assist the National Academies of Sciences, Engineering, and Medicine in making each published proceedings as sound as possible and ensure that it meets the institutional standards for quality, objectivity, evidence, and responsiveness to the charge. The review comments and draft manuscript remain confidential to protect the integrity of the process.

We thank the following individuals for their review of this proceedings: **Nadim Ahmad**, Organisation for Economic Co-operation and Development; **Thomas Guevara**, Indiana University; **Andrew Reamer**, The George Washington University; **Dominic Smith**, U.S. Department of Labor.

Although the reviewers listed above provided many constructive comments and suggestions, they were not asked to endorse the content of the proceedings nor did they see the final draft before its release. The review of this proceedings was overseen by **Evan Johnson**, University of North Carolina. He was responsible for making certain that an independent examination of this proceedings was carried out in accordance with the standards of the National

Academies and that all review comments were carefully considered. Responsibility for the final content rests entirely with the rapporteurs and the National Academies.

THE INNOVATION POLICY FORUM

The Innovation Policy Forum acts as a focal point for national and international dialogue on innovation policy. Operating under the guidance of the Board on Science, Technology, and Economic Policy, the forum brings together representatives of government, industry, national laboratories, research institutes, and universities—foreign and domestic—to exchange views on current challenges and opportunities for U.S. innovation policy. The forum provides a platform for learning about the goals, instruments, funding levels, and results of national and regional programs and for discussing their lessons for U.S. policy. The workshop described in this proceedings is one of a series of workshops and discussions convened by the Innovation Policy Forum since 2012.

Contents

1	INTRODUCTION	1
2	MULTINATIONAL FIRMS AND GLOBAL INNOVATION	7
3	TRACING VALUE ADDED IN THE PRESENCE OF MULTINATIONAL FIRMS WITH AN APPLICATION TO HIGH-TECH SECTORS	15
4	TRADE IN SERVICES, INTANGIBLE CAPITAL, AND THE PROFIT-SHIFTING HYPOTHESIS	23
5	TALENT, GEOGRAPHY, AND OFFSHORE R&D	31
6	THE NATURE AND DIRECTION OF INNOVATION IN GLOBAL VALUE CHAINS FOR WIND-ENERGY TECHNOLOGIES	39
7	ECONOMIES OF SCOPE AND RELATIONAL CONTRACTS: EXPLORING GLOBAL VALUE CHAINS IN THE AUTOMOTIVE INDUSTRY	47
8	KEYNOTE ADDRESS: FOREIGN DIRECT INVESTMENTS AND SUPERSTAR SPILLOVERS: EVIDENCE FROM FIRM-TO-FIRM TRANSACTIONS	55
9	CREATION AND DIFFUSION OF KNOWLEDGE IN THE GLOBAL FIRM	63
10	FIRM SELECTION AND ORGANIZATIONAL CHOICE: COMPLEX PATTERNS OF GLOBAL SOURCING	73

11	ARE CUSTOMS RECORDS CONSISTENT ACROSS COUNTRIES?	81
12	CAPITAL FLOWS IN GLOBAL VALUE CHAINS	89
13	COLOCATION OF PRODUCTION AND INNOVATION: EVIDENCE FROM THE UNITED STATES	99
14	GLOBAL VALUE CHAIN MEASUREMENT METHODOLOGY: CHALLENGES AND PROSPECTS	109
15	LESSONS FROM THE WORKSHOP: A PANEL DISCUSSION	125

REFERENCES 137

APPENDIXES

A	Workshop Agenda	153
B	Biographies of Speakers and Planning Committee Members	159
C	Crosswalk of Workshop Papers to Measurement and Understanding of Global Value Chains	173

1

Introduction

In recent decades, production processes of intermediate and final products have been increasingly fragmented across countries in what are called global value chains (GVCs). GVCs may involve companies in one country outsourcing stages of production to unrelated entities in other countries, multinational enterprises (MNEs) offshoring stages of production to units of the MNE overseas, or both. GVCs can also involve completely independent companies merely sourcing their parts from whichever upstream company may be the most competitive, with no control arrangement necessarily involved. The changing global trade environment and the changes in firms' behavior have raised new and more complicated issues for policy makers and have made it difficult for them to understand the extent and operations of GVCs and their spillover effects on national and local economies.

As providers of information to policy makers, researchers, and the public, national statistical organizations need to respond to the increasing prevalence of GVCs. However, while they publish a great deal of information on economic performance, trade, employment, and foreign direct investment (FDI), they have found it challenging to collect or present these data in a way that quantifies the international linkages and flows that characterize GVCs. National accounts are designed to measure economic activity within national borders. Asymmetries in accounting standards and data collection among countries further exacerbate the problems of gaining an accurate picture of economic activity and production, as well as market flows, within and across borders.

The increasing prominence of GVCs requires improved statistical frameworks, including those for business and trade statistics, to better reflect the increasingly complex economic activity that characterizes how goods and services are produced. Of particular importance is trade in research and development (R&D) services and other intangibles for which measurement is less straightforward and coordination across countries is necessary to produce comparable cross-country statistics. While the data for understanding trade in goods are robust, less is understood about trade in services, enterprise characteristics, and intangible assets. In the past, measurement was less complicated—production and R&D were conducted within the same firm, and a statistical agency could send the firm a survey. Now large firms produce and innovate in different enterprises and locations.

To improve the understanding, measurement, and valuation of GVCs, the National Center for Science and Engineering Statistics (NCSES) asked the Innovation Policy Forum at the National Academies of Sciences, Engineering, and Medicine to organize a workshop. A planning committee requested, reviewed, and selected papers to present at the workshop, which was titled "Innovation, Global Value Chains, and Globalization Measurement" and was held May 5–7, 2021. The research presented at the workshop built on a number of reports and conferences—involving the efforts of NCSES, the Organisation for Economic Co-operation and Development (OECD), and statistical organizations in many other countries over the past 15 years—on the intersection of globalization and intangibles and the challenges for official statistics. The project's full statement of task is in Box 1-1.

ROLE OF THE PLANNING COMMITTEE

This proceedings has been prepared by the workshop rapporteurs as a factual summary of what occurred at the workshop. The planning committee's role was limited to planning and convening the workshop. The views contained in the proceedings are those of individual workshop participants and do not necessarily represent the views of all workshop participants, the planning committee, or the National Academies of Sciences, Engineering, and Medicine.

OVERVIEW OF THE WORKSHOP

The workshop presentations summarized in this volume cover a variety of data sources from a variety of countries, as well as a number of policy questions. The goal for the presentations was to inform future research and data development and to help inform national statistical organizations and international collaborations working to improve measurement of R&D globalization. The workshop presentations focused principally on a series of papers invited by the planning committee and were presented by one or more of the authors of those papers, who in some cases also drew on their related work.[1] Appendix C describes how the papers contribute to better understanding GVCs, their measurement, and related policy issues. Additionally, the keynote address focused on challenges and prospects in GVC measurement methodology, and the panels discussed lessons learned from the workshop.

---

[1] References to presentations in this volume give the name of the presenting author(s). At times, they were joined in the session by additional coauthor(s), some of whom participated in question-and-answer portions of the session.

> **BOX 1-1**
> **Project Statement of Task**
>
> To inform the work of the Innovation Policy Forum, an ad hoc planning committee under the oversight of the Board on Science, Technology, and Economic Policy (STEP), in cooperation with the Committee on National Statistics (CNSTAT), will organize a two-day workshop to improve the understanding, measurement, and valuation of global value chains. A global value chain is a network of financially-independent companies located around the world involved in the production of a good or service and its global level supply, distribution, and post-sales activities. Drawing on the expertise of recognized experts on economics, industrial organization, trade, other social sciences; accounting and taxation of multinational enterprises; and statistics, along with the perspectives of personnel of national statistical offices, the workshop participants will:
>
> 1. explore efforts already under way to address gaps in national statistical systems with respect to the impact of global value chains;
> 2. offer perspectives on the highest priority efforts needed to improve supply- and value-chain data;
> 3. consider how to modify statistical systems to gather and incorporate the improved data; and
> 4. suggest ways to help the system respond to ongoing and future changes in the structure and organization of the economy.
>
> The committee will commission original research papers as background for the workshop. A proceedings of the presentations and discussions at the workshop will be prepared by a designated rapporteur in accordance with institutional guidelines.

Among the topics addressed in the presentations by invited paper authors were the relationship between production and innovation (Chapters 2, 5, and 13); the impact on trade of the treatment of capital through FDI (Chapter 3) or as an intermediate input (Chapter 12); and whether countries' tax policies affect intellectual property and trade measurement (Chapter 4). Two of the paper presentations dove deeply into a particular industry—wind energy (Chapter 6) and automobile components (Chapter 7).

A number of presentations focused on the issue of colocation of production and innovation. Gumpert (Chapter 2) described how she and her colleagues used data from Bureau van Dijk's Orbis database combined with information on intellectual property from the European Patent Office's (EPO's) PATSTAT database to investigate an MNE's choice of locations for production and innovation. They found that, compared with basic innovation, applied innovation is more likely to be colocated with production because of the synergies between applied innovation and production. They also found that basic innovation is more likely to be offshored to more advanced economies where human capital resources are likely to be more abundant, and that applied innovation is more likely to be offshored to emerging markets where production is cheapest.

Similarly, Fan (Chapter 5) described his use of the Orbis database and EPO's PATSTAT database to explore offshoring of R&D by MNEs, finding that the innovative intensity of an affiliate increases when the affiliate is located in a country with a higher-quality talent pool. He discussed his observation that innovation and production tend to be colocated.

Fort's presentation (Chapter 13) also explored colocation of production and innovation, based on her coauthored paper. Her presentation described the coauthors' look at U.S.-based companies using Census Bureau data to examine the relationship between manufacturing and innovation, specifically to determine whether colocating production near innovation increases innovation. While she and her coauthors found that colocating innovation near production does increase innovation, former manufacturing firms that no longer produce products continue to innovate. In fact, the share of patents from nonmanufacturers has grown over time and is almost half of all patenting activity.

Wei (Chapter 3) described his and his colleagues' use of OECD Analytical Activities of Multinational Enterprises (AMNE) data to account for the contribution of FDI in estimates of GVC activity. He showed estimates that do not account for FDI miss around half of what could be considered GVC activity, when accounting for value-added generated through FDI.

Similar to Wei, Ding's presentation (Chapter 12) described his use of tables from the World Input-Output Database (WIOD) to investigate whether treatment of capital inputs as an intermediate input affects trade statistics. He found that gains from trade are much larger when capital is considered an intermediate input.

Oddo's presentation (Chapter 4) described his and his colleagues' use of Italian tax data to determine whether there is mismeasurement of trade due to companies' shifting intellectual property products (IPPs) to tax havens. While they found strong evidence of such mismeasurement, their discussant cautioned that it is difficult to move IPPs and that such movements are usually overstated.

Two paper presentations dove deeply into particular industries. Surana and Doblinger (Chapter 6) described their and their colleagues' look at GVCs for wind technology using worldwide data from Navigant to find information on component manufacturing of wind technologies. Although previous researchers have found that outsourcing innovation has a deleterious effect on long-run innovation, Surana, Doblinger, and colleagues did not find such an effect; however, they did find a difference in geographic location of innovation depending on whether the firm was a component supplier or an original equipment manufacturer (OEM). They also found that component suppliers are likely to patent short- and long-term innovations regardless of whether they are in advanced economies or emerging markets, but that OEMs in advanced economies are more likely to hold long-term patents than OEMs in emerging markets.

In her presentation, Helper (Chapter 7) described her and her coauthor's use of U.S. customs data to explore firm-to-firm relationships in GVCs pertaining to the manufacture of components of automobiles. They explored whether firms organize their supply chains in response to organizational strategy. In such cases, firms may enter into purchase agreements not only to reduce transactions costs, but also to create tight-knit supply networks to build credibility and take advantage of economies of scope.

Tybout (Chapter 11) cautioned about the use of U.S. customs data. His presentation revealed that there may be issues in the accuracy of these data, although such microdata are essential to understand firm-to-firm transactions.

Warzynski (Chapter 10) described his and his coauthor's use of Danish data to investigate whether a firm will purchase from a firm in the same or another country. They used data collected by Statistics Denmark and surveys of offshoring coordinated with Eurostat. They found that while there has been an increase in outsourcing to the global South, this outsourcing has become more diffuse over time, with Danish firms outsourcing to firms in many more countries. They also found that MNEs are more likely to outsource to their affiliates.

Bircan's presentation (Chapter 9) described his and his colleagues' combination of geocoding from patent data from EPO, the U.S. Patent and Trademark Office, and the Japan Patent Office with Orbis data to explore the diffusion of innovation from MNEs. They found that barriers to diffusion may be related to time zone and physical differences. Gender also apparently matters—female investors appear to have less mobility than their male counterparts.

The keynote presentation at the workshop, by Van Reenen, focused on extensive spillovers to domestic firms from MNEs (Chapter 8), especially in high-R&D industries. According to Van Reenen, this work suggests the existence of a sales certification effect—that domestic suppliers that supply superstar MNEs tend to see increases in their sales, even outside of their sales to the MNE.

The penultimate session of the workshop (Chapter 14) focused on challenges for measuring GVC broadly, for modernizing official business statistics, and for measuring R&D and competitiveness for U.S. exports. The discussant (Francisco Moris) placed the research presented at the workshop within the context of extensive international efforts to develop and update guidance and standards for business and trade statistics, as well as measurement of R&D and intangibles. He proposed an integrated framework for identifying high-, medium- and low-priority projects to research, develop, and implement improved statistics on R&D, trade in services, GVCs, and relevant aggregates for the System of National Accounts.

In the final workshop session (Chapter 15), a panel was asked to identify lessons learned from the workshop. Individual panel members singled out the policy needs for statistics to understand the effects of GVCs on trade, different industries, and within-country economic growth or decline; the importance of finding innovative ways to share business data for national statistics and research, including linkage of microdata; and the use case for more detailed information on firm location and ownership, particularly for MNEs.

REMAINDER OF THIS VOLUME

The remainder of this proceedings volume is divided into chapters summarizing the keynote, paper presentations, and final panels in the order in which they occurred during the workshop. Chapters 2 through 13 summarize the presentations of workshop papers and the keynote. Chapter 14 summarizes a panel discussion focused exclusively on measurement issues, and Chapter 15 summarizes a final panel with reflections on the lessons derived from the workshop. These chapters are followed by a list of references that appear in the volume and three appendixes: Appendix A reproduces the workshop agenda, and Appendix B includes biographies of workshop speakers and members of the planning committee. Appendix C describes how the papers contribute to better understanding GVCs, their measurement, and related policy issues.

2

Multinational Firms and Global Innovation

Paper Authors: Anna Gumpert (Ludwig-Maximilians-Universität München [LMU Munich]), Kalina Manova (University College London), Cristina Rujan (Max Planck Institute for Innovation and Competition), and Monika Schnitzer (LMU Munich)

Presenter: Anna Gumpert (LMU Munich)
Moderator: Justin Pierce (Federal Reserve Board)

The workshop paper by Anna Gumpert, Kalina Manova, Cristine Rujan, and Monika Schnitzer explores the role multinational enterprises (MNEs) play in the diffusion of production and innovation across firms and countries. Offshoring of production decisions is a well-known phenomenon, and the authors extend their analysis to the innovation decisions of firms.

Anna Gumpert, assistant professor of economics at Ludwig-Maximilians-Universität München (LMU Munich) and research affiliate at the Centre for Economic Policy Research and at the Center for Economic Studies at the ifo Institute, introduced her presentation by observing how Mercedes-Benz and BMW, both German car manufacturers, can illustrate the process of innovation fragmentation and complex global value chains (GVCs). In 2017, Mercedes-Benz opened a research and development (R&D) laboratory in Seattle to focus on basic design innovation; and in 2018, BMW opened an R&D lab in Shanghai to focus on applied innovation, such as design and autonomous-driving technologies. These examples provide anecdotal evidence of MNE outsourcing of innovation activity to both developed and emerging economies. Gumpert argued that the examples of these firms are evidence of a larger global phenomenon. MNEs are the center for both global technological progress and the global fragmentation of production; they are also responsible for the majority of private R&D expenditures globally. These firms have also increased the fragmentation of their value chains across countries.

Gumpert argued that the contribution of the workshop paper lies in better understanding MNE innovation behavior, a field she said is little studied, outside of the papers discussed in this workshop.

More specifically, she indicated that their paper's contribution can be summarized as (1) providing novel facts based on a unique dataset of German MNEs, (2) offering an integrated model of MNE production and innovation, and (3) providing reduced-form empirical evidence consistent with the model assumptions and predictions. (Gumpert noted that this third aspect of the paper is still in progress.)

DATA

The paper uses data from Bureau van Dijk's Orbis database. These data, covering 15,000 German MNEs and their global affiliate networks from 1999 to 2016, are matched with firm-level data on patents over the same time span from the European Patent Office's PATSTAT database.[1] Gumpert observed that Germany is an ideal context for studying these questions as it is an innovation leader and is home to a large number of MNEs.

Gumpert described how these patent data are rich and offer several channels for inquiry.

- First, the data are characterized by inventor location, which allows the authors to conduct analysis based on domestic or offshore innovation location, and by whether the offshore innovation was colocated with offshore production. (Innovation and production are colocated if both activities occur within the same country.)
- Second, the data are characterized by type of patent. Patents are split into three categories: basic innovation, applied product innovation, and applied process innovation. An example of basic innovation is a new chemical reaction. Applied innovation is further from basic science. An example of applied product innovation is a new product based on the new chemical reaction, and an example of applied process innovation is a more efficient process for the new product. In the analysis by Gumpert and her coauthors, distance from science is determined by backward citations to scientific journals, and the different types of applied patents are assigned after a textual analysis of the patent abstract.
- Finally, the data are characterized by patent quality, which is measured by the number of forward citations of the patent 5 years after application.

---

[1] The authors originally intended to use data from Bundesbank's Microdatabase Direct Investment (MiDi). Although they were unable to access MiDi directly because of pandemic restrictions, Gumpert explained that they were able to use Bureau van Dijk's Orbis database instead to obtain the information.

STYLIZED FACTS

Gumpert presented three novel, stylized facts uncovered from the match between the firm data and patent data.

- First, MNEs innovate actively, and frequently abroad. Nearly one-third of German MNEs own at least one patent. Nearly one-third of patent-holding MNEs have a patent with a foreign inventor. Therefore, 15 percent of those in the entire MiDi database are foreign inventors.
- Second, German MNEs offshore innovation to locations with and without an affiliate present. More than half (56 percent) of German MNE innovative activity is done only at home, and very few MNEs innovate solely abroad whether or not they have a foreign affiliate. At the same time, 14 percent of MNEs innovate in all locations (at home, offshore where they have an affiliate, and offshore where they don't have an affiliate).
- Third, the type and intensity of innovation varies across firms, but firms that hold more patents tend to hold patents of significantly higher quality.

MODEL

Gumpert presented an integrated model of MNE production and innovation strategy. This model assumes

- heterogeneous firms with respect to productivity,
- three countries (home and two host countries),
- a constant elasticity of substitution (CES) demand curve with no trade costs, and
- three types of innovation (basic, applied product, applied process).

Countries differ on multiple dimensions. The production wages are assumed to be highest in the home country and lower in the two hosts, with the hosts having different wages. The authors impose no structure on inventor wages. They assume that fixed costs of innovation are equal in the two host countries and that the fixed cost of innovation is lowest in the home country. The three types of innovation are captured in the model by three cost functions. Basic innovation will increase future profits, applied product innovation decreases the fixed cost of adding a new product, and applied process innovation decreases the marginal cost of production. The cost functions are similar in that endogenous innovation costs increase with innovation quality.

Firms choose the location and intensity of innovation to maximize global profits. The paper authors allow for more generality in the full paper, but for the purposes of her presentation Gumpert focused on the case in which production location is given, innovation only occurs in one location, and the only innovation type is basic. Innovation can take place in conjunction with production in host 1 or without production in host 2.

Gumpert then presented the full maximization problem for this simplified model. Firms choose the innovation intensity of basic innovation, q, to maximize their profit function, pi. The production side assumptions are standard (CES demand) and contain variable profits and simple fixed costs. This means that basic innovation either increases future profits or decreases the exit probability of the firm. In the paper, this maximization is done on all types of innovation. When firms innovate, they face innovation costs that are the sum of a fixed innovation cost and the variable cost of innovation that is increasing in quality. The variable costs of innovation also depend on colocation of innovation and production. For applied innovation, the variable costs of innovation are lower if innovation is colocated.

Gumpert noted that, while the paper includes seven testable predictions, she focused on just four of them in her presentation:

- First, more productive MNEs are more likely to innovate—and innovate more intensively—in each innovation type. This result is due to the structure of the profit function being supermodular with respect to productivity and innovation quality.
- Second, more productive MNEs are more likely to offshore innovation of each type. This result is driven by higher fixed costs of offshore innovation relative to domestic innovation.
- Third, applied innovation is more likely than basic innovation to be colocated with production. There are synergies between production and applied innovation that do not exist for basic innovation.
- Fourth, MNEs are more likely to offshore basic innovation to countries with a comparative advantage in innovation, and applied innovation to countries with a comparative advantage in production. This is driven by the same synergies that drive prediction three.

Gumpert also briefly mentioned the remaining predictions. Fifth, the elasticity of the extensive and intensive margins of innovation with respect to firm productivity may vary across innovation types. Sixth, more productive MNEs are more likely to offshore innovation in house. Lastly, more productive MNEs are more likely to offshore basic innovation at arm's length to countries with a comparative advantage in innovation, and applied innovation in house to countries with a comparative advantage in production.

EMPIRICAL EVIDENCE

Although she noted that the empirical section of the paper is still in progress, Gumpert presented some of the early results.

- The authors found significant and positive results suggesting that larger firms are more likely than smaller firms to patent at all, to patent more, and to produce higher-quality innovation. This supports the first prediction.
- They also found significant and positive results suggesting that larger, or more productive, firms have more foreign innovation by all types of patents. This result is strongest for applied process patents. The results also suggest that these firms have a larger share of foreign innovation relative to smaller firms across all innovation types. This supports their second prediction.

Gumpert and colleagues also tested a corollary of the model, that offshore innovation is of higher quality, and found that in all specifications, the foreign-only patents are of higher quality than the coinvented patents. These results are all weakly significant. They also found that applied innovation is more likely than basic innovation to be colocated with production, the excluded categorical variable in the results table. This supports their third prediction.

Gumpert emphasized that these results are all conditional correlations and are not to be interpreted as implying causal relationships. In order to test for causality, they studied the impact of distance to high-quality universities. They found significant and negative results suggesting that firms that innovate more are physically closer to all universities. However, after controlling for firm size, only the distance to the top university is significant. Gumpert added that she and her colleagues plan to explore this last relationship, as well as the relationship of university quality, in future work.

SUMMARY

Gumpert closed by pointing out key takeaways. First, MNEs innovate intensively and frequently abroad, offshoring innovation to both production and nonproduction locations. Second, more productive MNEs innovate more intensively than less productive MNEs, and they are more likely to offshore innovation. Lastly, offshore applied innovation is more likely than offshore basic innovation to be colocated with production. Gumpert noted that avenues for future work include identifying the optimal innovation and tax policy in developed countries and the optimal foreign direct investment policy for developing countries seeking to attract capital and technology.

DISCUSSION

Discussant: Allison Derrick
(Bureau of Economic Analysis, U.S. Department of Commerce)

Allison Derrick, of the Bureau of Economic Analysis (BEA) in the U.S. Department of Commerce, began her discussion of the paper by briefly summarizing the work. The authors' main objective, she said, is to understand how MNEs produce and innovate. Three types of innovation—basic, applied product, and applied process—are presented and incorporated into a formal model that has several predictions. The model suggests that the different types of innovation have different incentives, and preliminary empirical work using firm-level patent data shows that some predictions of the model appear to be supported.

Derrick indicated that the main contribution of this work is using MNE data to shed light on an open empirical question: how do MNEs innovate? She noted that the dataset used, Orbis, relies on public data and will not be as in-depth or reliable as the confidential MiDi database proposed, and she suggested that the results may or may not hold in the preferred dataset. The paper contributes to the literature by understanding where different stages of innovation take place. This has important implications for developing countries and trade.

Derrick noted that, although it is young, the literature on MNEs and innovation is established and provides several relevant papers. Data from the BEA, which are similar to MiDi data, have been used to study similar questions. The BEA data have also been linked to patent data in a similar fashion to that of Gumpert and her coauthors. For example, Bilir (2014) studied how intellectual property (IP) rights influence MNE production decisions; Berry (2014) found that multicountry patents have grown over time; and Branstetter and colleagues (2019) documented the globalization of R&D and the rise of new R&D hubs, and observed that the increase in foreign R&D among U.S. MNEs is driven by software and information technology industries.

Derrick expressed major concern about the external validity of the results; specifically, whether patenting activity varies across industry and location or the results are driven by the data in the sample. Patenting variation is driven by variable R&D costs and imitation costs, life-cycle length of products (incentive to not apply for a patent until a clear use is identified), and IP protection in the producing/innovating location. Derrick suggested that Gumpert and colleagues should control for patent differences by industry and location. This has the potential to reveal important patterns in the model while also checking to see if one or two industries are driving the results. Without mentioning specifics, she recommended robustness checks for these results.

Derrick suggested that there are several key insights from a paper she is currently working on with her colleague Christopher Steiner. In that work, they looked only at innovation with firm boundaries and have not studied the different stages of innovation. Their work has been motivated by the increases in R&D service imports and exports over the last two decades. Derrick indicated that she is surprised by the large number of German MNEs that collaborate with unaffiliated parties, which is quite different from patterns among U.S. MNEs.

Derrick also mentioned other work relevant to the subject paper. Previous research has shown that the measurement and observation of innovation are difficult (see Corrado et al., 2017; Lev and Gu, 2016). There are three proposed measures of innovation: R&D expenditure and employment, IP-related services, and patents. Derrick and Steiner (in progress) use the first two categories, but not patents. They found that 85 percent of R&D is conducted at home and about 60 percent of R&D-intensive MNEs are in manufacturing. Derrick stated that there are two kinds of innovators. First are adapters, those that conduct applied product and process innovation to fulfill production roles. Adapters are common but small in size, and most belong to a manufacturing MNE. Second are R&D centers and IP sellers, those that perform basic or applied innovation and for which R&D may be their primary industry. Relative to adapters, both R&D centers and IP sellers report more sales, have higher R&D expenditures, trade more IP services, are more common among nonmanufacturing MNEs, and are more productive. The major takeaway of this, she said, is that Gumpert and colleagues should look at how industries drive their results because firm innovation behavior is heterogeneous along industry and type.

Another concern Derrick raised is the measure of productivity. The subject paper uses log global sales to measure productivity. A concern with this as a measure is that it directly measures the size of the firm, which may be correlated with firm productivity. However, she said that this concern may be alleviated by access to the MiDi dataset, which has better measures of firm productivity, such as value-added productivity.

Derrick also suggested that patent ownership is likely affected by tax rate differences. As shown in Derrick and Steiner (in progress), the majority of R&D takes place in nonhaven countries, while ownership and royalty payments are shifted to tax-haven countries. In relation to the subject paper, there is the question of how many of the foreign inventors live in tax havens.

Finally, Derrick offered some miscellaneous comments and questions. First, how is "arm's length" determined? Does this mean there is no affiliate with a German parent or no affiliate within the entire MNE? This is important as the German parent may not be the global owner. Second, for innovation at arm's length, with whom are the MNEs innovating? This may help shed light on the first causal relationship between innovation and distance to a top university. Lastly, some descriptive statistics on patents should be included.

Justin Pierce, principal economist at the Federal Reserve Board and moderator of the session, commented that the authors should decompose innovation quality in a similar fashion to innovation location.

Anna Gumpert thanked Allison Derrick for a great discussion and the many helpful comments and suggestions. While several suggestions were given, she offered two main responses.

First, in response to the questions Derrick posed about measures of innovation, Gumpert clarified that in their paper, she and her colleagues used only patents owned by German parent firms. This should alleviate some of the concerns raised about tax havens and external validity. R&D expenditure is not a useful measure in this case as it does not allow for analysis of foreign inventors; rather it picks up on the flow of money.

Second, addressing Derrick's suggestion that the results may not be robust to the MiDi data, Gumpert indicated that this is not a concern because the correlations are also present in the MiDi data, although the exact coefficients are different.

During the presentations, one audience member asked about the implications of the use of trade secrets in applied product or process innovation, specifically innovation that isn't picked up in patents, in firms with foreign affiliates. In response, the authors offered that, while it is impossible to parse this out as they are trade secrets, this would be a larger concern with process innovation because firms are less likely to patent process innovation than product innovation. Another audience member asked where the foreign innovation is taking place and if it is driven by countries like India or China; the authors responded that their work shows that most foreign innovation takes place in countries neighboring Germany.

3

Tracing Value Added in the Presence of Multinational Firms with an Application to High-Tech Sectors

Paper Authors: Zhi Wang (George Mason University and University of International Business and Economics), Shang-Jin Wei (Columbia Business School, National Bureau of Economic Research [NBER], and Centre for Economic Policy Research [CEPR]), Xinding Yu (University of International Business and Economics), and Kunfu Zhu (Renmin University)

Presenter: Shang-Jin Wei (Columbia Business School, NBER, and CEPR)
Moderator: Sally Thompson, Bureau of Economic Analysis (retired)

The workshop paper by Zhi Wang, Shang-Jin Wei, Xinding Yu, and Kunfu Zhu explores the accurate measurement of global value chains (GVC) activity, which is vital to the validity and success of future international trade research.

Shang-Jin Wei, the N. T. Wang professor of Chinese business and economy and professor of finance and economics at Columbia University's Graduate School of Business and School of International and Public Affairs, introduced the presentation by describing the goals and features of the paper. The paper develops a new accounting framework to better account for the role of foreign direct investment (FDI) in GVC accounting systems, consistent with the national accounts conceptual framework. The paper offers two accounting decompositions, which collapse to a single formula in matrix notation: a forward and backward linkage.

The paper's major contribution is the addition of FDI-related activity to a GVC accounting framework. The authors found that previous estimates of GVC activities that do not account for the role of foreign-invested firms miss about half of GVC activities. Specifically, the authors found that GVC activity involving foreign-invested enterprises (FIEs) accounts for about 10 percent of global gross domestic product (GDP), similar in magnitude to the entirety of GVC activities identified in current approaches used to measure GVC activities using conventional input-output tables (that do not separately identify the contribution made by MNEs and foreign affiliates). These additional GVC activities are larger in high-tech sectors and higher-income countries.

The measurement framework in this paper decomposes a country's GDP and final production into different value-added terms related to pure domestic activities, trade-related GVC activities that involve trade in intermediate inputs but not through FDI, and FDI-related GVC activities. The value of their approach is illustrating the importance of FDI in determining participation in GVCs and, in turn, informing bilateral investment treaties.

Wei indicated the importance of accounting for FDI by showing two data patterns:

- First are the trends in FDI stock and trade volumes. Both are generally increasing with time, but the global FDI stock has increased at a quicker and growing rate. Both are related to GVC activity, so why are they increasing at different rates?
- Second, about 60 percent of U.S. multinational corporation–affiliate sales in 2016 were in the host country of the affiliate. This pattern displays the importance of host-country sales in the GVC.

PAST LITERATURE AND CONTRIBUTION

Wei began the review of the literature on GVC accounting with Hummels and colleagues (2001), who introduced two measures of trade in intermediate goods. Koopman and colleagues (2012) updated this framework to include domestic value added. Noguera (2012) proposed a net measure of value added in trade, but did not capture the entirety of domestic value added in exports. Koopman and colleagues (2014) next proposed a framework containing four mutually exclusive terms that account for all gross exports. The framework accounts for domestic value added that is exported and consumed abroad, domestic value added that is exported and then imported, foreign value added used in the production of exports, and double-counted items.

Following the work of Koopman and colleagues (2014), there are two strands of literature. The first strand studies the decomposition of gross trade. Wei cited the following papers without additional comment: Nagengast and Stehrer (2016), Johnson (2018), Borin and Mancini (2019), Arto and colleagues (2019), and Miroudot and Ye (2020). The second strand of literature studies the decomposition of value added beyond trade. Wei cited Los and colleagues (2015), who decomposed final good production but made no distinction between GVC and non-GVC production activity. Wei also cited Wang and colleagues (2017), who decomposed GDP and final good production in a fashion that measures GVC activity.

According to Wei, this workshop paper contributed to the second strand of literature with three major contributions: (1) adding to the literature on measuring GVC activity using intercountry input-output (ICIO) tables by studying the contribution of FIEs, (2) contributing to literature on FDI in global production fragmentation and GVCs, and (3) quantifying FDI-related GVC activity by country income and industry research and development (R&D) intensity.

ACCOUNTING FRAMEWORK

Wei briefly discussed the underlying methodology. The authors started from ICIO tables with information on firm ownership. There are G economies, N industries, and two ownership types. This creates ICIO tables in which firm behavior is split by ownership, separating the total output of foreign- and domestic-held firms. The decomposition framework follows Wang and colleagues (2017), the major contribution of which is a single-equation decomposition "$\hat{V}L\hat{Y}$." This decomposition is extended in the workshop paper to include FDI.

For simplicity, Wei focused on forward-linkage decomposition only. The decomposition containing FDI includes 16 terms that can be characterized into five categories: pure domestic production activity, traditional trade production activity, trade-related GVC activity, FDI-related GVC activity, and trade- and FDI-related GVC activity. Wei claimed that the FDI-related GVC activity is missed in current estimates of GVC activity that do not fully account for the contribution of value added generated through FDI channels to GVCs.

Wei explained that the 16 terms can be compressed into nine components. These nine components are split into non-GVC activity and GVC activity. Three of the components are classified as missing GVC activity. The missing activity centers on FIE production and value added for final goods consumed in the host country.

DATA AND NUMERICAL APPLICATION

The workshop paper uses the Analytical Activities of Multinational Enterprises (AMNE) database from the Organisation for Economic Co-operation and Development (OECD). This database has the form of an ICIO table broken down by domestic- and foreign-owned firms. The dataset includes 59 countries and a rest-of-the-world composite and covers 34 industries from 2005 to 2016.

The paper shows that the largest contribution to global GDP is made by pure domestic activity, which appears to be countercyclical. The remainder of the global GDP can be expressed in the four other categories. FDI-related GVC activity is the largest nondomestic component of global GDP. Trade-related GVC activity is the next largest, contributing about 2 percent less to global GDP per year than the FDI GVC contribution. Non-GVC trade accounts for about 5.5 percent of global GDP across all years; this is about 4 percent less than the FDI-related GVC activity in a given year. The trade measures of GVC activity—pure and hybrid trade and FDI—shrank during the global financial crash in 2008. The GVC activity related to FDI did not have a strong response to the financial crash, and Wei suggested this may be an interesting channel for future macroeconomic research.

The results Wei presented are heterogeneous across countries in the sample. FDI-related GVC activity plays a larger role in smaller and more open economies, such as those of Hong Kong, the Czech Republic, Singapore,

Romania, and Hungary. In such countries as South Korea, Russia, Japan, Saudi Arabia, and Israel, the FDI-related component of GVC activity contributes less to GDP.

Wei explained that variation due to industry is also present in the sample. For this analysis, countries are split into low, middle, and high income, and industries are split into medium-low, medium, and high tech. Two patterns arise in the data:

- First, for high-income countries, the GVC-activity share of value added is highest in high-tech industries and lowest in low-tech industries. This result is robust across the three income groups. This trend is also true when considering only the FDI-related share of GVC activity.
- Second, in high–R&D-intensity industries, FDI GVC activity is more important than trade-related GVC, whereas in medium- to low-tech industries, the GVC components are equal.

SUMMARY

Wei concluded by summarizing the paper's main results. According to Wei, the paper develops a framework to trace value added that accounts for FIEs. He indicated that the framework allows for both forward and backward linkage; however, he focused his discussion on the results from the forward-linked decomposition. He and his coauthors use the Analytical AMNE database from the OECD to study how the introduction of FDI-related GVC activity alters the existing estimates of the share of GVC activity in global GDP, finding that prior methods miss 9–10 percent of GVC activity contributing to global GDP. After correcting for this missing activity, the total contribution of GVC activity to global GDP is about 20 percent—double the prior estimates. There are some country and industry patterns to the missing data. The volume of missing activity is largest in high-tech industries and high-income countries, relative to medium-tech industries and middle-income countries. There is no formal hypothesis testing in this paper but the decomposition results provide a more accurate measurement of GVC activity that can be used in future work. Finally, Wei concluded that their future work may include incorporating bilateral FDI stock into the accounting framework, as well as decomposing gross exports.

DISCUSSION

Discussant: Thomas F. Howells III (Bureau of Economic Analysis, U.S. Department of Commerce)

Thomas Howells, chief of the Industry Analysis Division of the Bureau of Economic Analysis (BEA) in the U.S. Department of Commerce, began his discussion by acknowledging the contributions of the paper and providing some background information to contextualize it in the current literature on measuring GVC activity. Howells stated that the paper is well placed in the literature and builds on the use of trade in value added (TiVA) statistics. The backward-linkage approach would take a final product and look backwards at its supply chain, whereas the forward-linkage approach would look at a given industry and analyze GVC activity in a forward direction. The literature typically offers a decomposition of either exports or final demand. Howells explained that this paper contributes by developing a framework containing both forward and backward linkages, as well as a decomposition of final demand. A decomposition based on exports is listed as future work by the authors.

Howells observed the importance of the ICIO tables to the literature, noting that this paper uses an expanded ICIO, including multiple other dimensions for variation—namely domestic- and foreign-owned firms. To help understand the main contribution, which is the framework itself, Howells presented a simple example of product flows that the framework would capture. In this example, a U.S. firm purchases intermediates from a domestic and a foreign firm. This U.S. firm may then trade intermediate goods back to these firms and/or may sell final products directly to consumers, both domestically and internationally. In this setting, U.S consumers may buy final products from the foreign firm as well.

Howells pointed out that standard TiVA analysis defines GVC as cross-border trade in intermediate goods and services, breaking trade into three categories: domestic production and consumption, pure trade of final goods and services, and GVCs. While the implied assumption in this framework indicates that a firm's location is what matters, the paper posits that firm ownership is also an important dimension of analysis.

Howells described the empirical results presented in the paper as interesting and important to future work. The main takeaway from the empirical results is that current TiVA analysis misses about 10 percent of the FDI-related GVC activity in terms of global GDP.

Howells offered several comments and critiques. The first comment concerned the Analytical AMNE database from the OECD used in the empirical analysis. The framework itself is robust to choice of dataset, but the empirical results are not. The standard OECD ICIO tables contain around 6 million cells, while the Analytical AMNE database contains approximately 24 million cells. While data exist for many of these cells, the OECD did have to construct additional data, and doing so introduced assumptions into the data. The OECD team used tools like gravity modeling, ordinary least squares regression analysis, and linear interpolation. Howells noted that the task the OECD attempted was very difficult and the team did a fantastic job.

Howells moved to a discussion of the BEA research agenda. The BEA is developing extended supply-use tables, which will contain information on three types of firms: U.S. parents, U.S. affiliates, and an "other" category. The BEA is pursuing a top-down approach, using aggregated and readily available datasets, as well as a bottom-up approach, using extensive microdata. A National Bureau of Economic Research (NBER) working paper characterizes the work on the top-down approach.[1]

Howells then offered some final comments:

- First, the standard TiVA framework provides a breakdown of value added by labor and capital location. In the new framework, control of assets is an important dimension. Howells differentiated between control and ownership, and noted foreign minority stakes, which are larger than 10 percent in BEA data and may have sizable control of a firm, especially where the minority stake is a plurality of shares. In order to be considered an FIE, a firm must have 50 percent foreign ownership in the OECD database.
- Second, the new framework can indicate that a firm in country C manufactured an intermediate part assembled in country B and purchased in country A. The framework cannot indicate the location of the parent of the firm in country C. This means that the framework is still somewhat location locked, like the standard TiVA framework.
- Third, clarification is needed in how we think about the GVC decomposition terms outlined in the chapter. Rather than being mutually exclusive, we can consider the total value created in any GVC as coming, in different amounts, from the different components.
- Lastly, the standard TiVA framework accounts for trade in intermediate goods and services, but MNEs will also trade capital goods. This is an interesting future path for work.

Wei agreed that Howells made a useful distinction between FIEs by ownership and FIEs by control. If this is a concern, then the estimates are undercounting the role of FIEs in their empirical analysis. Wei also indicated that the conceptual framework is independent of the dataset used and encouraged its use with whatever data are available.

Finally, Wei indicated that current data do not allow for removing the location-based aspect of GVC, as discussed in the above example with three countries. Wei expressed hope for this data to become available, but he said that he and his colleagues are working on an alternative approach to answer this question.

---

[1] According to Howells, the NBER paper breaks down value-added content in exports for a set of industries in 2012 and shows that value-added content in exports varies widely across industries. U.S. MNEs appear to have large market share in the high-tech R&D-intensive industries, while U.S. non-MNEs have larger market share in manufacturing industries. Imported content is concentrated in the high-tech R&D-intensive industries, as is foreign MNE domestic value-added content. See Fetzer et al. (2021).

Nadim Ahmad, deputy director at the OECD Centre for Entrepreneurship, SMEs, Regions and Cities, added that the connection between the TiVA framework and the FDI flows was the most intriguing part of the presentation. This connection is vital to truly understand how GVCs work and was a large motivator for the OECD in the creation of its expert group on extended supply use tables, which it formalized in 2014. Ahmad cautioned that the presented framework is important but may go too far by quantifying any FDI activity by an MNE as GVC activity.

Wei responded that there is a distinction between simple and complex GVC activity and not all FDI-related GVC activity is treated as the same in the current framework.

4

Trade in Services, Intangible Capital, and the Profit-Shifting Hypothesis

Paper Authors: Nadia Accoto (Bank of Italy), Stefano Federico (Bank of Italy), and Giacomo Oddo (Bank of Italy)

Presenter: Giacomo Oddo (Bank of Italy)
Moderator: Nadim Ahmad (Organisation for Economic Co-operation and Development)

The workshop paper by Nadia Accoto, Stefano Federico, and Giacomo Oddo explores the profit-shifting actions of multinational firms (MNEs) in the context of intellectual property products (IPPs). Like the workshop paper presented by Shang-Jin Wei (Wang et al., Chapter 3), this paper discusses potential mismeasurement in trade in services by looking at intangible capital, such as IPPs. The concern motivating this paper is that international service payments to related parties can be strategically manipulated to reduce tax obligations by creating income in low-tax jurisdictions and deductions in high-tax jurisdictions.

Giacomo Oddo, an economist at the Bank of Italy, introduced his presentation by defining some important terms. IPPs are immaterial and exchangeable goods or assets, such as patents, trademarks, copyrights, software, managerial expertise, algorithms, databases, and research and development (R&D) information. Payment for these products and services qualifies as trade in services and enter balance-of-payment statistics. About one-fifth of total services trade in the European Union is composed of trade in IPP or intangible assets. IPPs have an increasing presence on the balance sheets of MNEs (Haskel and Westlake, 2018), and these products are easily and cheaply transferable across borders and between firms (Griffith et al., 2014). Together, they provide MNEs new strategies for profit shifting.

Oddo explained that the paper uses Italian firm-level data to understand this phenomenon. The analysis contains three steps: (1) analyze the geographic and industry characteristics of Italy's IPP services trade, (2) apply the Tørsløv methodology (Tørsløv et al., 2018) to Italian firm-level data, and (3) verify whether profit-shifting estimates and imports of IPP services are correlated at the firm level.

Oddo continued that the paper relates to two vast strands of literature: the rising role of intangible capital in the economic activity of firms (Corrado et al., 2009; Haskel and Westlake, 2018; Jenniges et al., 2019; Jona-Lasinio and Manzocchi, 2012) and the profit-shifting activity of MNEs to tax havens (Barrios and d'Andria, 2020; Bilicka, 2019; Bruner et al., 2018; Clausing, 2016; Davies et al., 2018; Dharmapala, 2014; Riedel, 2018; Sallusti, 2019; Tørsløv et al., 2018). The paper contributes to the literature at the intersection of these two strands of work by focusing on the role of intangible capital in profit shifting. Other work in this area includes Dischinger and Riedel (2011), Griffith et al. (2014), Beer and Loeprick (2015), Alstadsæter et al. (2018), and Barrios and d'Andria (2020).

DATA AND DESCRIPTIVE STATISTICS

Next, Oddo stated that he and his coauthors used a dataset that merges Bank of Italy data on trade in services with balance sheet data, and covers 2,600 Italian firms from 2013 to 2017. There are 30 types of services, following the extended balance of payments services (EBOPS) classification, which the authors aggregated into three types: IPP services, headquarter (HQ) services, and other services. The IPP services category contains royalties for the use of and rights to intellectual property, software and computing services, and general R&D behavior and information. The HQ services category contains accounting, auditing, and tax services; managerial and entrepreneurial consulting; and other services between associated firms. The other services category includes residual services in the dataset not included in the prior types. Both the IPP and HQ service types allow for profit shifting as they contain services in knowledge-based sectors.

The paper follows Hines and Rice (1994) and Tørsløv et al. (2018) and divides countries by their status as a tax haven. Oddo presented descriptive statistics, based on his and his coauthors' sample of 2,600 Italian firms, about the distribution of trade in services by their tax-haven status. IPP services imports come from tax-haven countries at a higher proportion with respect to HQ or other services. There are also differences across sectors. Manufacturing firms play an important role in the international trade of services—when looking at trade in IPP services, these firms account for 67 percent of exports and about 40 percent of imports. Exports of IPP services are concentrated relative to imports of IPP services, for which there is more dispersion in activity.

Oddo explained that IPP services are traded overwhelmingly by very large firms (defined as those with more than 1,000 employees), which account for more than 70 percent of export activity and approximately 63 percent of import activity. The authors also considered ownership status, dividing the sample into foreign-owned firms (those whose parent company is located abroad), local firms (those whose parent company is in Italy), and firms that are not part of any group. Close to 60 percent of IPP service imports are made by foreign-owned firms. Oddo pointed out that one interesting dimension of variation is the difference between ownership and control brought up in the discussion of the workshop paper presented by Shang-Jin Wei (Wang et al., Chapter 3). With some regression analysis, the authors found that foreign-owned firms are positively correlated with IPP trade activity after controlling for size, industry, and year fixed effects.

METHODOLOGY

Oddo described the Tørsløv et al. (2018) method as comparing the profitability rates of local and foreign-owned firms. Profitability is determined by an index that divides the pretax corporate profit by the compensation of employees. The central finding of the Tørsløv et al. (2018) paper is that firm profitability is higher in tax havens for foreign-owned firms, while firm profitability is higher for local firms in non–tax havens. Assuming all firms face Cobb-Douglas production functions, meaning an assumption of equal capital intensities between foreign-owned and local firms, then a nonzero difference between local and foreign-owned profitability must be due to profit shifting. Shifted profits are then the difference between expected profits (profits at the same profitability index as local firms) and their actual profits.

Acknowledging that this methodology is imperfect, Oddo presented four potential weaknesses, which all reflect data limitations:

- First, there is an implicit assumption that local firms do not engage in profit shifting and therefore their profitability is the true profitability.
- Second, the Cobb-Douglas assumption (stating that there are equal capital intensities between foreign-owned and local firms) may not hold empirically.
- Third, the definition of foreign-owned firms does not align between the Foreign Affiliates Statistics and FDI data employed in this method.
- Lastly, the depreciation of foreign-owned firms is obtained as a residual and may lead to implausible estimates for some countries.

By applying the methodology to microdata, the authors overcame the third and fourth issues, so their efforts were focused primarily on the first two issues.

EMPIRICAL RESULTS AND APPLICATION TO ITALIAN FIRMS

Oddo said that he and his coauthors first verified the firm profitability index results from Tørsløv et al. (2018), namely that foreign-owned firms are less profitable than local firms in Italy, presenting significant results that confirm this relationship. The authors aggregated microlevel data in three ways to compute the profit-shifting activity of foreign-owned firms:

- First, the authors followed the past literature by aggregating across the entire sample of firms, known as the direct approach.
- Second, they aggregated on an industry-by-industry basis and then used a process known as the sum-across-sectors approach.
- Lastly, they aggregated on a firm-by-firm basis, comparing each foreign-owned firm with the average of local firms in the same industry; this is known as the granular approach.

The second and third aggregation techniques offered similar results, and Oddo focused on the results of first two before turning to the third technique later in the presentation.

Oddo introduced several results from analyzing profit shifting in all industries at once (see Figure 4-1). According to analysis using the first aggregation technique—the direct approach—foreign-owned firms shift 32 percent of adjusted profits. According to the second aggregation technique—sum-across-sectors—an estimate of 15 percent of total adjusted profits are shifted by foreign-owned firms; however, the profit-shifting intensity varies with industry. The importance of industrial composition is highlighted by the difference in the two estimates. The macro approach used in Tørsløv et al. (2018) may introduce nonnegligible bias to estimates (see Barrios and d'Andria, 2020). Relaxing the Cobb-Douglas assumption, employing instead a range of values for the elasticity of substitution between labor and capital from 0.7 to 1.3, alters the profit-shifting estimates to between 4 percent and 42 percent of adjusted profits, respectively. Both approaches offer similar values to other estimates in the literature. The macro estimate, 32 percent, is lower than Tørsløv et al. (2018), while the micro estimate, 15 percent, is very close to the estimate in Sallusti (2019), which uses the granular approach on Italian firm-level data.

Oddo offered a comparison of the shifted profits by foreign-owned firms with the value of IPP (and HQ) services imported by the same group of firms. This analysis is broken into two parts: a macro and micro stage.

- The macro stage asks if the IPP (and HQ) import flows are large enough on aggregate to accommodate the predicted macro profit-shifting estimates. The profit-shifting estimates using the macro approach are too large to be accounted for with only imports of IPP and HQ services from tax havens.
- The micro stage asks if profit-shifting firms are the same firms that import IPP and HQ services. The more conservative micro approach yields an estimate of shifted profits that can be explained almost completely by the IPP and HQ imports from tax havens. However, there is still some difference in the two values, implying that all imports of IPP and HQ services from tax havens were for the purpose of shifting profits.

- The 1st approach (direct) quantifies the size of shifted profits as **32%** of adjusted profits, while the 2nd approach (sum across sectors) points to a lower amount: **15%**.
- The discrepancy between the two approaches is relatively large, suggesting that **sectoral composition matters**, and that macro approach like Tørsløv et al. (2018) may cast a non-negligible bias on estimates (Barrios and D'Andria, 2020).
- Releasing Cobb-Douglas assumption, considering tangible capital intensities from balance sheet data, and assuming σ to be in the range 0.7–1.3, our **estimates vary between 4% and 42% of adjusted profits**.
- Our "macro" estimate (32%) is lower than what Tørsløv et al. (2018) (however, close to their estimate with adjusted depreciation), while our "micro" estimate (15%) is very close to that of Sallusti (2019) (13%), who estimates profit shifting through a granular approach on Italian firm-level data.

FIGURE 4-1 Profit-shifting estimates.
SOURCE: Presentation by Giacomo Oddo.

In the second stage of their analysis, Oddo described, the authors asked if profit-shifting firms are the same firms that import IPP and HQ services from tax havens. The authors used the third approach to aggregation, outlined above, and correlated the estimated shifted profit with the firm-level flow of imported IPP and HQ services. At this level, shifted profits are correlated only with imports of IPP from tax havens. This means that the HQ services do not appear to be related to profit shifting. This relationship is stronger when considering only the subsample of large importers of services. This subsample of firms displays high rates of imports of IPP services from tax havens and has large alleged shifted profits.

SUMMARY

Oddo concluded by discussing the contribution of the paper to the current literature. The authors used both macro- and micro-level approaches to estimate the profit-shifting activity of firms with respect to imports of IPP services. The authors find that Italian imports of IPP services are compatible with the hypothesis that such flows are used for the purpose of shifting profits to tax havens. Forty percent of imports of IPP services come from tax havens, while only 30 percent of imports of other services come from tax havens, with large foreign-owned firms accounting for almost 66 percent of this activity. The baseline estimates of shifted profits vary between 15 percent and 30 percent of adjusted profits; however, the estimates are dependent on modeling assumptions that may cast doubt on the results. Lastly, Oddo reported that he and his colleagues found a positive correlation at the firm level between profit shifting and IPP imports from tax havens; but they also found that, at the aggregate level, other channels also contribute to profit shifting.

DISCUSSION

Discussant: James Hines (University of Michigan)

James Hines, Richard A. Musgrave collegiate professor of economics and L. Hart Wright collegiate professor of law at the University of Michigan, began the discussion by complimenting the authors on good work, great data, and their contribution to the literature on an important and fascinating topic. Hines expressed confidence that firms shift profits using IPP service trade as a channel. However, he argued that the current estimates in the literature greatly exaggerate the extent to which this occurs. Hines described this paper as offering great data for further study of profit-shifting behavior of firms, but thinks a thorough investigation of the literature is important.

Hines continued by noting the difficulty present in attempting to use service transactions to shift income. Due to rules on arm's-length pricing, an MNE cannot simply shift intangible assets to a foreign affiliate in a tax haven. Rather, the foreign affiliate must acquire the asset, which creates taxable income in the jurisdiction where the asset changes hands. Royalties on the intangible assets are supposed to take place at a fair market rate; however, MNEs have an incentive to understate the value during the transfer and to overstate the value when paying royalties. This means it is very difficult to measure what the genuine market prices should be, which provides reasons for most companies to avoid shifting intangible assets to tax havens—still, some do engage in this activity. Hines reiterated that there are rules and the rules are enforced, but he also acknowledged that the rules are much more difficult to enforce for trade in services.

Hines brought up the concern that IPP services are the main channel for shifting profits from high-tax countries to low-tax countries, which, given the difficulty in valuing such services, is well founded. One major concern around the estimates of profit-shifting activity is that the counterfactual is unknown; in other words, it is unknown where the income would have been earned if there were no havens. Another major concern is the double-counting of profits in tax havens. For example, if a U.S. firm invests in a German affiliate and routes it through Bermuda, then the German profits show up in the Bermuda profits, even though the German profits were taxed in Germany. Hines explained that these prior critiques do not necessarily pertain to this paper, but rather to the literature as a whole. Moving to a more relevant critique for this paper, Hines observed that employment or labor compensation is a bad proxy for the location of where profits are truly earned, as firm profits are decreasing in labor expense (Hines, 2010); thus, this proxy is a valid option only if there are no alternatives.

Hines asserted that, while it is true that MNEs shift profits and arrange financing to minimize their tax burden, it is difficult for them to do so; only the largest firms engage in this activity. Fewer than half of U.S. MNEs and fewer than 20 percent of German MNEs have any tax-haven affiliates. Additionally, if such profit shifting was easy to do, there would not be a negative correlation between tax rates and FDI. If profit shifting was easy, MNEs would not care where their FDI went since they would shift the profits to a haven regardless of the tax rate in the invested country.

In order to show that profit shifting is overreported, Hines presented an example from Germany, where firms have a strong incentive to locate profits in low-tax countries, because of high German taxes. However, the data show that only a small number of firms, 20 percent, have an affiliate in a tax haven. For the largest German MNEs, those that report annual revenue of at least 759 million euros, this percentage is higher, but even these firms reported less than 9 percent of their 2016–2017 profit as earned in a tax haven. On the other hand, the Tørsløv et al. (2018) methodology yields an estimate that 64 percent of German corporate-sector profits is shifted to tax havens, a value that is impossibly high given the 9 percent figure above.

Moving forward, Hines suggested that the authors focus in on more subtle, disjointed questions. Most important, when and why do firms engage in this behavior? Within this question, what observable factors, in addition to firm size and industry, are associated with tax-haven activity? Many firms report negative profits, and it would be interesting to understand what role IPP payments and receipts play in the reported profits. Do firms that pay IPP royalties to tax-haven affiliates also receive royalties from them, or are they unaffiliated firms? What are the observable factors associated with Italian affiliates that receive royalties for IPP when the payer is an affiliate firm in a tax haven? Lastly, to what extent do patterns of service payments differ based on the nationality of foreign-owned firms? Hines suggested Dharmapala and Riedel (2013) and Clausing (2001) as empirical models that may aid in this research. He ended by noting that this is an exciting project and that he is looking forward to the authors' future work.

Oddo began by thanking Hines for the enlightening and delightful discussion. He agreed with Hines with respect to the downsides of the Tørsløv et al. (2018) methodology and with the assessment that the estimates in the literature are exaggerated. He thanked Hines for the valuable suggestions for future work.

During the question-and-answer segment, moderator Nadim Ahmad, of the Organisation for Economic Co-operation and Development, suggested that the authors look at differential tax rates over time to explore temporal changes. Hines also suggested that, given that profits vary over time and many companies experience losses in some years, the authors examine whether there is a correlation between higher overall profitability and royalty payments to tax-haven affiliates.

5

Talent, Geography, and Offshore R&D

Paper Author and Presenter: Jingting Fan (The Pennsylvania State University)

Moderator: Eduardo Morales (Princeton University)

Jingting Fan, assistant professor of economics at The Pennsylvania State University, introduced his presentation by providing the motivation behind his new model of trade and innovation. Global economies have become more integrated because of increasing trade volumes and the activities of multinational enterprises (MNEs). There is a growing body of literature on the costs and benefits of globalization. Past literature has focused on offshore production of MNEs; in contrast, this paper focuses on offshore research and development (R&D), defined as the fraction of R&D that occurs in a host country by an affiliate firm of a foreign corporation. Fan's workshop paper extends prior models of trade and R&D to enable exploration and understanding of the patterns of the global diffusion of R&D.

Fan explained that offshore R&D, which is R&D performed by affiliates of foreign corporations within a country relative to total domestic R&D activity, is the majority or substantial share of total R&D activity. Smaller economies are more reliant on R&D expenditures by these affiliates of foreign firms than are larger economies and, in general, R&D performed by affiliates of foreign firms is growing over the sample period (1985–2012). For example, in the United States, by the end of the sample period, offshore R&D increased from about 7 percent to 15 percent of total U.S. R&D (see Figure 5-1).

Fan described two components required for innovation and commercialization of new products: a talented workforce and know-how of firms. But he said there is a spatial mismatch between these two components. Countries like China and India have some of the largest talent pools, while the vast majority of well-run firms are in highly industrialized countries. Thus, offshoring R&D can have a direct impact by helping with this spatial mismatch, as well as an indirect effect on countries through trade and offshore production. There are several challenges to analyzing the impact of offshoring R&D:

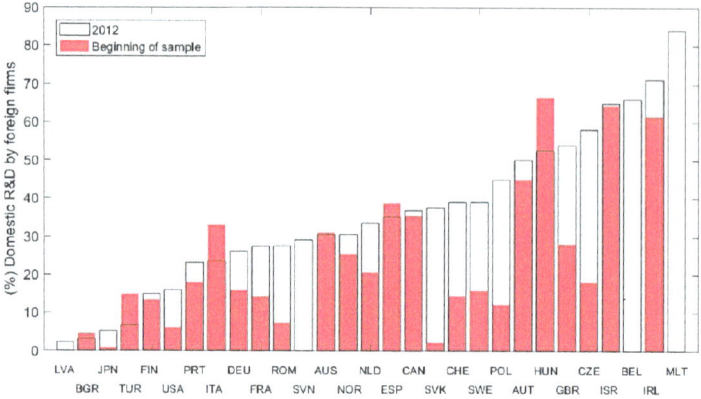

FIGURE 5-1 Large and growing offshore research and development (R&D).
NOTE: OECD = Organisation for Economic Co-operation and Development.
SOURCE: Presentation by Jingting Fan.

- First, R&D and production are jointly determined and depend on market access.
- Second, calibrating the model to fit realistic geographic features requires firm-level data on the activities of MNEs in different regions, and such data are relatively scarce. Additionally, most firm-level data focus on one home country at a time.

Fan stated that his paper contributes to the literature by assembling a panel dataset that merges production and ownership data from Bureau van Dijk's Orbis database with patent data from the European Patent Office's PATSTAT database. Using this panel dataset, the paper documents several empirical relationships, including the role of human capital in affiliate R&D, colocation of affiliate R&D and production, and headquarter (HQ) effects for both affiliate R&D and production. These patterns are then interpreted and explored using counterfactual experiments in a structural model. There are two proposed mechanisms within the model: talent acquisition and market access. Used first as a measurement tool and second as a counterfactual experiment, the model produces the following results:

- First, around 70 percent of R&D in overseas affiliates is for domestic production, and offshore R&D is an important source of profit for firms headquartered in advanced economies. This

indicates that the market-access motive is a strong incentive for firms to innovate abroad.
- Second, the movement from zero offshore R&D to positive offshore R&D generates 3.3 percent welfare gains on average and amplifies total gains from globalization by 33 percent. These estimates have significant bias from advanced economies, as the advanced economies benefit more from offshore R&D. There are also important interactions between offshore R&D and trade and offshore production.

DATA

Fan continued by stating that he merged data from the Orbis database with patent data from PATSTAT covering 37 countries with four periods over 1996–2016. Parent firms in the Orbis database are defined as an entity with at least 50 percent control of a firm. Affiliate behavior in a foreign country is the aggregate of all firms controlled by the parent in the given country. Patent behavior was matched to firms before being aggregated to three dimensions: parent firm, inventor country, and time. The paper accounts for a large amount of this variation by controlling for firm, industry, and host fixed effects and their interactions, in addition to other factors. The estimates of the model align well with expenditure data from the Organisation for Economic Co-operation and Development (OECD).

Fan moved on to describe the structure of the data. Here he explained an inference problem due to the difficulty in assigning invention location for given production. For example, the innovating location may be the home country for production in a host country, or vice versa. There are more dimensions to this problem, and the picture can quickly become quite complicated. To work around this inference concern, he uses the data to document a systematic distribution of innovation activity. Then, knowing the size and location of production, one knows the thickness, or size, of the flows of innovation going to each node in the network. Fan then parametrized the value of each link in the network, under some assumptions.

STYLIZED FACTS

Fan presented four stylized facts:

- First, firm heterogeneity plays an important role, as firms with more HQ innovation also have higher affiliate innovation in the extensive and intensive margins, and the affiliate firms have higher sales per invention.

- Second, there is a positive correlation between host human capital and affiliate innovation intensity, or the ratio of patents and sales, in cross-section and panel dimensions.
- Third, there are synergies to the colocation of innovation and production. For example, firms that conduct innovation in one host country are also more likely to produce in that country. This relationship is true when focusing on the intensive margin and changes over time.
- Lastly, there are benefits to being geographically close to the firm HQ, as both affiliate innovation and production decrease with distance to HQ.

MODEL

Fan moved next to discussing the model environment. The model is nested in the work of Arkolakis et al. (2018), with the inclusion of offshore R&D as the major change. There are N countries endowed with L workers who have ability (alpha), drawn from a distribution A. Workers either work in manufacturing and earn a common wage w, or work in a high-skill job and earn a premium wage (w * alpha). Firms are heterogeneous and are differentiated in their manufacturing productivity and their efficiency at innovation. A representative consumer maximizes a constant elasticity of substitution (CES) demand function, subject to a budget constraint.

The firm's decision begins at the firm HQ in the home country (see Figure 5-2). The parent firm chooses to invest in R&D in two host countries. The affiliate firms can then choose to produce and sell in either the host country or another destination market. Fan provided the example of DuPont, a U.S. MNE with R&D labs in the United States, Brazil, China, Switzerland, South Korea, Germany, and Japan. DuPont produces in 19 countries around the world and sells final products in close to 90 countries.

Fan explained that firms that wish to invest and innovate in a host country pay a fixed entry cost. These firms retain some of the innovation efficiency gained from the parent firm, which is located in the home country, depending on the distance between the home and host countries, and gain a production efficiency drawn from a distribution. Firms can enter multiple host countries at once, incurring multiple fixed costs and drawing a unique productivity efficiency in each. Firms face a unit delivery cost of production determined by the retained productivity efficiency, the production wage, the manufacturing productivity, and the distance between the host country and the parent firm, as well as the distance between the host firm and the final market.

Fan stated that firms also incur a marketing cost in the final market and a shipping cost from the production location to the final market. For technical

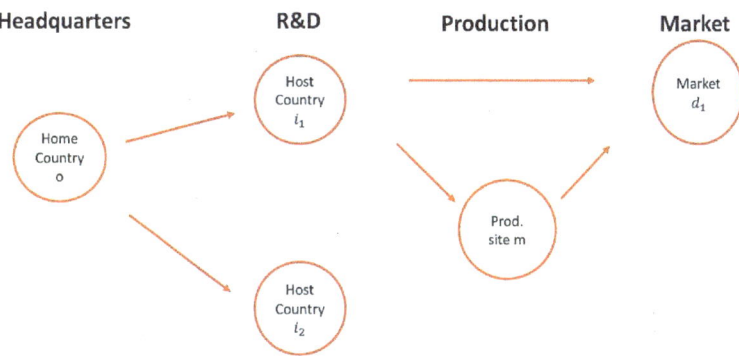

FIGURE 5-2 Firm decision.
NOTE: R&D = research and development; HQ = headquarters.
SOURCE: Presentation by Jingting Fan.

reasons, there is an idiosyncratic component that varies across regions. The revenue generated by the final sale of products will be split across the value chain, leaving the net profit with the HQ.

The model includes horizontal innovation, which Fan described as new product blueprints/inventions that are differentiated from each other. Inventions are produced at a cost that takes into account both the distance between the production and R&D (via wage premiums and production efficiency), as well as between production and the final market (via shipping and marketing costs).

To calibrate the model to the real world, Fan explained, the paper uses country size as a proxy for manufacturing employment. For the distribution of talent, the paper follows Hanushek and Woessmann (2012), which is based on a country's PISA score (from the OECD's Programme for International Student Assessment) and the distribution of cognitive skill. For the firm know-how distribution, the World Management Survey from Bloom et al. (2012) is utilized. The production efficiency is the average of target, operations, and monitor scores, and the innovation efficiency is based on the talent score.

Fan modeled offshore production as a weighted average of the distance between the final market and the firm's HQ, and the distance between the final market and the production site with host-specific parameters. This is similar to the approach used in Arkolakis et al. (2018), but with the inclusion of the offshore R&D centers. Offshore R&D costs are modeled as a function of host-specific parameters and a distance measure between the home and host countries. The host-specific parameters measure the overall openness to foreign production and R&D for both the intensive and extensive margins. The paper does this by taking the share of foreign-owned firms that either produce or innovate in a host country. The distance measure is weighted by a coefficient from a regression measuring the HQ effect and benefits for colocation. A prediction of this model is that the proximity to the R&D center is more important for decreasing production costs than proximity to the HQ. Fan noted that extensive literature is available on the importance of HQ effects, but these results suggest that the proximity to the site of innovation is much more important.

RESULTS

Fan then transitioned to discussing the main results of his study. About 65 percent of domestic R&D is completed by domestic firms, and about 35 percent is conducted by foreign-owned firms, on average. The model predicts that the share of R&D activity for the purpose of production in the host country is higher among domestic than among foreign-owned firms. As shown in Figure 5-3, emerging economies have a higher share of R&D dedicated to domestic production than developed economies, both for domestic and foreign R&D. Foreign R&D contributes about 1.5 percent of total profits; however, there is large heterogeneity within foreign R&D contributions. For example, foreign R&D contributes 7.6 percent of total profits in the United States, whereas in Brazil, foreign R&D contributes 0.01 percent of total profits.

Fan also presented several counterfactual experiments. Offshore R&D generates about 3.3 percent welfare gains on average, which are larger for

Country	Source and use of R&D				Source of income (% of total income)				
	% by domestic firms		% by foreign firms		mfg.	profit		R&D	mkt.
		% of local prod.		% of local prod.		total	inventions abroad		
	(1)	(2)	(3)	(4)	(5)	(6)	(7)	(8)	(9)
RUS	88.4	91.7	11.6	81.6	79.1	9.4	0.06	3.2	8.3
BRA	42.4	97.4	57.6	84.1	76.6	10.6	0.01	4.5	8.3
CHN	58.5	95.0	41.5	74.3	76.7	10.2	0.00	4.8	8.3
MEX	45.0	98.9	55.0	94.7	76.9	10.7	0.01	4.1	8.3
TUR	79.9	97.9	20.1	89.3	75.4	12.0	0.00	4.2	8.3
...									
DEU	71.8	70.5	28.2	59.4	76.9	11.2	1.69	3.6	8.3
FRA	73.4	71.0	26.6	55.8	76.4	11.6	1.39	3.7	8.3
GBR	37.0	46.8	63.0	39.0	83.9	5.4	1.99	2.4	8.3
BEL	41.1	63.6	58.9	45.9	85.0	4.6	0.72	2.0	8.3
USA	84.2	67.6	15.8	58.3	66.0	20.8	7.65	4.9	8.3
Mean	65.6	82.9	34.4	69.5	78.3	10.0	1.5	3.4	8.3

- Column 6: profit share of total income; Column 7: offshore R&D profit share of total income
- Average value of Column 7 is 1.5%, highly heterogeneous across countries

FIGURE 5-3 The input-output of offshore research and development (R&D).
SOURCE: Presentation by Jingting Fan.

developed countries. The introduction of offshore R&D amplifies the total gains from liberalization by 1.3, or a 30 percent increase. Offshore R&D acts as a substitute for trade and offshore production for developing countries, and as a complement for developing countries. The combination of general equilibrium effects and firm linkages leads to the conclusion that incorporating offshore R&D is crucial when evaluating trade and offshore production policies.

SUMMARY

In conclusion, Fan stated that his work calibrates a model of trade, offshore production, and offshore R&D, studying the determinants and welfare implications of offshore R&D. According to Fan, the inclusion of offshore R&D is a novel contribution, building on recent literature and noting several new stylized facts from the merging of the Orbis and PATSTAT databases. The theory-based measurements show that the empirical patterns are rich and informative, and the counterfactual experiments highlight the importance of offshore R&D for understanding the implications of globalization for welfare and income distribution. Finally, he concluded with a caveat, that his work includes no industry-level analysis and overlooks the role of outsourcing in both production and R&D.

DISCUSSION

Discussant: Gary Lyn (Iowa State University)

Gary Lyn, assistant professor of economics at Iowa State University, began his discussion by summarizing Fan's work as a theoretical model consistent with three empirical facts: (1) the innovation intensity of an affiliate firm increases in a host country with a higher-quality talent pool; (2) innovation and production tend to be colocated; and (3) geography matters for the location of affiliate production and innovation. The model contains two motives for the observed trends in offshore R&D: talent acquisition and market access. Fan also conducted several counterfactual experiments to quantify the gains from R&D offshoring.

Lyn offered three broad comments on Fan's work:

- First, the introduction in Fan's paper would be enhanced by a discussion of the relationship of this work to Arkolakis et al. (2018), as the latter describes innovation as occurring only in the HQ country, and finds that countries that specialize in production may be hurt by offshoring R&D. This is related to the counterfactual in which no innovation efficiency is transferred. Additionally, in Fan's paper, the variety-level labor productivity is drawn from a multivariate distribution. Arkolakis et al. (2018) allows for labor productivity to be correlated across draws. Allowing such correlation may offer interesting predictions.

- Second, related to work on R&D spillovers, Lyn cited Bilir and Morales (2020), which describes HQs as experiencing stronger spillovers to overseas affiliates than spillovers from one affiliate to another. In other words, there appear to be HQ effects for knowledge transfer. Lyn asked if there is an easy way to implement this behavior in Fan's model. He suggested associating scale effects with employment in R&D in the home country, and then interacting these scale effects with the HQ innovation efficiency. Lyn's discussion prompted an audience member to ask if there are empirical R&D spillovers from foreign affiliate firms to foreign nonaffiliate firms.
- Third, Lyn raised a question about the broader story being presented: Is there a systematic difference in the type of R&D activities of HQs and affiliates? In other words, are affiliates innovating on different parts of the value chain to make a single product? This relates to the paper presented by Anna Gumpert (Gumpert et al., Chapter 2) that breaks down R&D activities by basic, applied product, and applied process innovation, and presents some stylized facts about these relationships. In addition, Lyn wondered why vertical foreign direct investment is not a first-order segment of the narrative.

Fan began his response with a comment about the relation of his work to Arkolakis et al. (2018), saying that labor productivity correlation is already captured in the firm-level productivity draws. This means that all varieties of R&D will be produced with a common level of efficiency and generate firm-level correlation.

Second, regarding the difference in innovation types, Fan returned to the work of Gumpert and her coauthors as a path forward.

Third, regarding the spillovers between affiliate firms in different hosts, Fan pointed out a positive correlation between production and distance to affiliate R&D centers. As an example, an affiliate R&D center in China will have some influence on an affiliate production center in India. This implies a scope for cross-country within-firm spillovers. Fan said he has not thought about the within-country spillovers across different affiliates, but thinks it is an important angle for future work.

During the presentation, one audience member asked about how the model handles the relationship between sales and R&D expenditures in the context of HQ and affiliate firms. Fan responded that this would appear in the model when innovation occurs in the host country and production occurs in surrounding countries. Eduardo Morales of Princeton University asked a question about the impact of human-capital accumulation and the firm's decision to offshore innovation, suggesting that the PISA score may not be the best measure. Fan responded that, while the PISA score is not perfect, it is the best measure available.

6

The Nature and Direction of Innovation in Global Value Chains for Wind-Energy Technologies

Paper Authors: Kavita Surana (University of Maryland), Claudia Doblinger (Technical University of Munich), Deyu Li (University of Cambridge), Nathan Hultman (University of Maryland), and Laura Diaz Anadon (Cambridge Centre for Environment, Energy and Natural Resource Governance, and Harvard Kennedy School)

Presenters: Kavita Surana (University of Maryland) and Claudia Doblinger (Technical University of Munich)
Moderator: Eduardo Morales

      Kavita Surana, assistant research professor at the Center for Global Sustainability in the School of Public Policy at the University of Maryland, and Claudia Doblinger, of the Technical University of Munich, presented their workshop paper, authored with Deyu Li, Nathan Hultman, and Laura Diaz Anadon. Unlike the papers already presented, Surana et al. uses global value chains (GVCs) to understand innovation in one particular technology: wind energy. Current predictions of global carbon dioxide emissions indicate that, in order to meet long-term decarbonization goals, a sizable increase in innovation in sustainable energy will be required. Technologies that are currently at the prototype or demonstration stage today are estimated to contribute 35 percent of the necessary reductions in 2070, and a further 40 percent will come from technologies that are currently at early adoption stages.
      Surana stated that the direction of innovation in clean-energy technologies depends on the location of production centers in the GVC. She and her coauthors developed a temporal outlook for technology introduction and advances, focusing on when the final products will begin to go to market and when they will be needed. The production choices of multinational enterprises may change the direction of innovation; for example, manufacturing shifts to Asia may reduce U.S. and global innovation in the short and medium term, as was seen in automobiles and high-end optoelectronics (Fuchs, 2014). As documented in Binz and Truffer (2017) and Pietrobelli and Rabellotti (2011), production and innovation are dispersed globally through the GVC. This leads to a main research

question: How does the location of suppliers in the GVC impact the direction of innovation?

The authors focused on the wind-energy industry, in which technology innovation occurs at the component level, leading to larger and better wind turbines. The literature on the wind-energy GVC focuses on three dimensions:

- First, on the turbine technology itself. For example, when turbines increase in size, they become more efficient.
- Second, on the leading firms or the original equipment manufacturers (OEMs)—globally, there are 15.
- Third, on the policies of countries, particularly the transition of innovation from the European Union and the United States, to China and India.

While all this work is important, Surana explained that a more subtle and in-depth review of wind turbines indicates a highly complex value chain with many innovative component suppliers.

CONTRIBUTIONS AND RESULTS

Surana stated that she and her coauthors have made several contributions, some of which are completed, and some of which are still in progress:

- First, they mapped the GVC for wind-energy technologies.
- Second, they analyzed the location of producers, both for supplying firms and the large-scale OEMs.
- Third, they measured and analyzed the direction of innovation and the connection to long-term societal goals.
- Lastly, the authors analyzed the relationship between location and innovation in the GVC.

Surana described these contributions in greater depth in the order above. To measure and map the wind-energy GVC, the authors took a bottom-up approach, as there were no available datasets to answer their questions. They looked at wind-energy reports published every 2 years by a private research firm, Navigant Research, analyzing the wind-energy supply chain. The authors have information on 389 suppliers of nine components representing more than 1,000 relationships with 13 OEMs between 2006 and 2016. This period is limited but does correspond with large amounts of innovation in the wind-energy industry. There are a few countries that supply a large majority of the turbine components. These facts were established by linking supplier firms to their location, size, founding year, and mergers and acquisitions, and to whether they specialize in the wind-technology industry. Surana encouraged the interested reader to look at this work, which is already published (Surana et al., 2020).

Surana and her coauthors used turbine towers as an example to discuss how the location of producers—both suppliers and OEMs—impacts the wind-energy industry. They found that component suppliers are working increasingly with OEMs across international borders. In 2006, the suppliers were located in countries with large OEMs and large markets for wind energy. Between 2006 and 2016, suppliers emerged in new locations, especially in countries without an OEM. Over this period, the number of suppliers increased by a factor of 15, and countries with a supplier increased by a factor of 3. New suppliers, located in Africa, Latin America, and Asia Pacific, concentrate production on low-complexity parts, such as towers and generators. New suppliers of high-complexity parts, such as blades and gearboxes, are rare (see Figures 6-1 and 6-2).

Surana then shifted to a discussion of the long- and short-term trends in innovation activity. The authors measured the direction of innovation by studying the content of patents in the space. There are more than 12,000 patents from suppliers and OEMs in the dataset, which the authors clustered into component type, technology, and type of improvement they address. This is done by understanding the cooperative patent classification, sorting the patents into 40 categories by their type, and by studying the key terms in the text of the patent. The authors chose to use these assessment tools instead of looking at patent citations. The authors matched wind-energy needs identified by the International

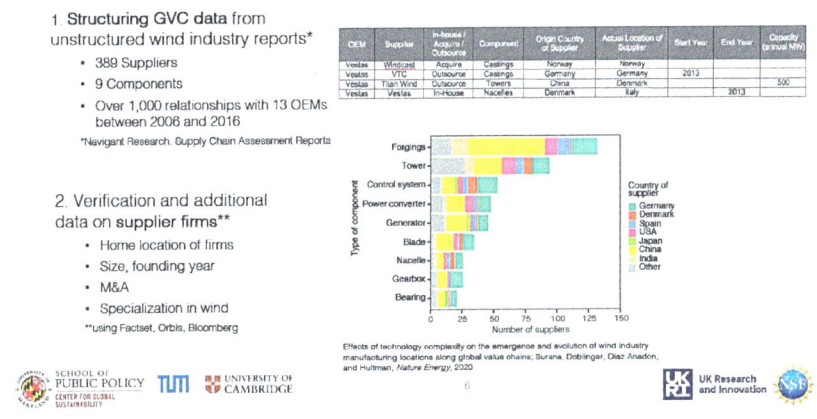

FIGURE 6-1 Measuring and mapping the wind-energy GVC.
NOTE: GVC = global value chain; M&A = mergers and acquisitions; MW = megawatts; OEM = original equipment manufacturers.
SOURCE: Presentation by Kavita Surana.

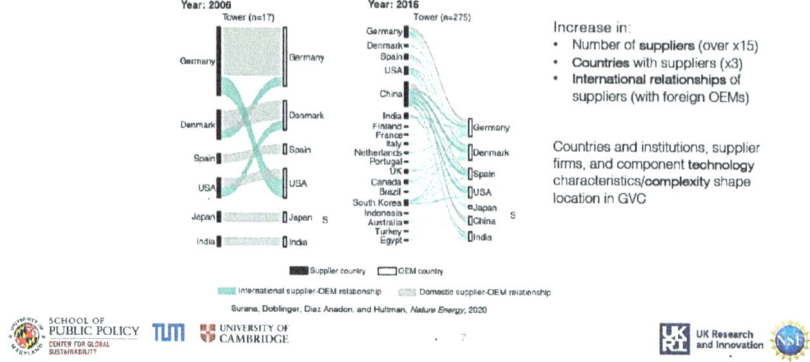

FIGURE 6-2 Analyzing the location of manufacturing: Component suppliers increasingly work with OEMs (e.g., tower suppliers) from other countries.
NOTE: GVC = global value chain; OEM = original equipment manufacturers.
SOURCE: Presentation by Kavita Surana.

Energy Agency (IEA) to the text of patents, allowing them to characterize the innovations as short term—dealing with installation, assembly, and adapting to environment conditions—or long-term, such as offshore wind technologies.

At this point, Claudia Doblinger, assistant professor for innovation and technology management at the Technical University of Munich, began presenting. Doblinger continued the discussion of the impact of suppliers and OEMs on the direction of innovation. The paper organizes results by producer type, by the stage of the host economy (either emerging or advanced), and on the temporal dimension. The temporal dimension (short, medium, and long term) was found by identifying needs from the reports on wind research and development (R&D) from the IEA and mapping them back to the patent content. For example, short-term patents include installation and assembly, while long-term patents relate to offshore wind technologies. Long-term innovation, both by component suppliers and OEMs, is concentrated in advanced economies; however, emerging economies, such as China, India, Brazil, and Mexico, are increasingly innovating in this dimension. Both OEM and component suppliers are innovating, with about 40 percent of new patents filed by component-supplying firms. Component suppliers have a similar proportion of short- and long-term patents across economy type, while OEMs in developed countries have a higher share of long-term innovation—about 50 percent across the sample—compared with OEMs in emerging economies, which had a share of about 25 percent in 2016.

Doblinger presented regression coefficients from analysis on shifts in market growth and production in emerging economies, primarily China. The authors found that shifts in GVC activity to emerging economies did not reduce long-term innovation activity, which is dependent on local market size and growth. Local relationships are defined as an E.U. supplier collaborating with an E.U. OEM, or a Chinese supplier and a Chinese OEM. International relationships are defined as a non-E.U. supplier and an E.U. OEM, or vice versa, or a Chinese supplier and any non-Chinese OEM, or vice versa. The results are split between before and after 2012, which marks a significant shift in the production and market for wind-energy technology from the European Union and other developed countries to emerging economies, especially China. The results suggest that the increased globalization of GVCs in wind-energy technology did not decrease long-term innovation in the European Union after 2012, in contrast with the findings by Fuchs (2014) for automobiles and high-end optoelectronics. International relationships linked to market demand are more important than the origin country of the supplier for long-term innovation.

SUMMARY

A key takeaway, according to Doblinger, is the importance of the temporal dimension to innovation in this industry. Further, this work contributes to the literature by studying the entire wind-energy GVC at both the supplier and OEM levels, which is important for understanding policies that may increase innovation in this industry. The authors found that shifts in innovation from the global North to the global South did not decrease long-term innovation, which may be true for other industries. Doblinger said she and her coauthors hope to further refine their method of analysis and to implement formal hypothesis testing to further understand the direction of innovation in the future. They also hope to assess and understand the importance of regional demand for wind energy in China. Finally, they hope to trace innovation spillovers between OEM firms and supplying firms, as well as the impact these spillovers may have on long-term innovation.

DISCUSSION

Discussant: Kelley Sims Gallagher (Tufts University)

Kelley Sims Gallagher, professor of energy and environmental policy, director of the Climate Policy Lab, and codirector of the Center for International Environment and Resource Policy Institute in the Fletcher School at Tufts University, began her discussion by stating that she considered the authors' findings interesting, and that the main motivation for her questions and comments is to understand why these patterns exist and what mechanisms may lead to these results. In her prior work, Gallagher examined how China has taken advantage of globalization to develop production and innovation capacity in clean-energy technologies, particularly in solar technology.

Gallagher suggested that a possible driver of China's emergence in the clean-energy GVC is its export-oriented production strategy and its global perspective on markets. This is partly driven by the fact that Chinese students study globally and notice market opportunities in other countries. China was also investing in infrastructure, both physical and human, which may best explain the findings in the workshop paper. In addition, trade liberalization and Chinese entry to the World Trade Organization in 2000 likely contribute to this effect.

Gallagher noted that in Figure 6-1, China and Germany appear to be the only countries with firms supplying all nine components. She wondered why these countries are succeeding compared with other countries, despite market share changes over time, and whether there are implications for industrial policy. A possible reason may be that China and Germany are more interventionist in industrial policy relative to other countries.

Gallagher added that in Figure 6-2, which displays the temporal dimension of suppliers in the market for wind turbine towers, there is an increase in suppliers from emerging economies. She wondered whether the authors have considered this shift in suppliers responsible for any cost reductions resulting from increases in competition. Another policy question is the importance of local content requirements in driving the fragmentation of the GVC over the decade between 2006 and 2016.

Gallagher found interesting the results on short- and long-term innovation by suppliers and OEMs differentiated by emerging or developed economies. She referred to the investment and industrial policy of China as a leading hypothesis for the uptick in long-term innovation by the emerging economies and wondered whether the authors can offer any ideas for why there is a peak in patenting behavior, for both short- and long-term innovation in 2010. Lastly, she asked if the authors have considered whether China's industrial policy may be preventing a decrease in long-term innovation, despite the fragmentation of the GVC.

Surana agreed with Gallagher that Chinese industrial policy may be preventing a decrease in innovation, but also offered another factor. Domestic Chinese demand for clean-energy technology likely plays an important role in the direction of innovation. The authors presented the splits pre-2012 and post-2012 to offer some evidence of this trend. They are working on integrating domestic industrial policy, among other policy decisions, into their paper.

Addressing questions on why China and Germany seem to dominate the market share of supplying firms in the clean-energy GVC, Surana referenced Surana et al. (2020), in which she and her coauthors attempted to answer this question. Part of the deviation can be explained by the complexity of the component.

Doblinger addressed the questions about the role of Chinese industrial policy preventing a decrease in long-term innovation. The breakdown of local and international relationships indicate that this is not the entire story.

Although the authors have not done a large amount of work on the 2010 peak in patenting behavior, Doblinger believed it is an interesting path for future work. Deyu Li, research associate at the Centre for Environment, Energy, and Natural Resource Governance at the University of Cambridge and a coauthor of the presented work, added that the decline in patenting after 2010 may be driven by the low number of countries active in the industry. The authors did not study the number of active firms in the industry, but literature suggests that, over time, the number of firms innovating in both solar and wind technologies has decreased. International firms also benefited from their connection to Chinese firms, and their long-term innovation increased after connecting with a Chinese firm.

Surana agreed that looking at the cost side of the fragmentation in the GVC is interesting.

During the panel question-and-answer period, a question from the audience focused on the competitiveness of the market at global and domestic levels and what is the typical consumer. Surana offered that the market is becoming more competitive with time, and the typical consumer is a private firm establishing a wind-energy farm. A second question concerned the trends in long-term innovation, particularly on whether China's entry into wind-energy markets increased or decreased long-term innovation in the United States. Doblinger responded that this was the main motivation for the project. They did not find a decrease in long-term innovation in the wind-energy sector, but offer no evidence for other industries.

7

Economies of Scope and Relational Contracts: Exploring Global Value Chains in the Automotive Industry

Paper Authors: Susan Helper (Case Western Reserve University) and Abdul Munasib (Bureau of Economic Analysis)

Presenter: Susan Helper (Case Western Reserve University)
Moderator: Wolfgang Keller (University of Colorado Boulder and National Bureau of Economic Research)

The workshop paper by Susan Helper and Abdul Munasib explores the development of supply networks by considering multiple transactions at a time. The authors report that transaction-based measures are significant and organization-based measures generally have much greater explanatory power.

Susan Helper, the Frank Tracy Carlton professor of economics at the Weatherhead School of Management at Case Western Reserve University, introduced her presentation with a story about the Kojima Press. In the late 1930s, Hamakichi Kojima asked Toyota for work, eventually obtaining a contract for procurement of sand buckets for use in fire protection. Toyota trained Kojima and, after some time, ordered additional automotive parts for a truck radiator. Now, Kojima supplies a wide variety of metal and plastic automotive parts to Toyota through a global network of affiliates. This story offers several puzzles, particularly from the point of view of transaction cost analysis.

Helper explained that Toyota often offers suppliers a "permanent deal," giving long-term contracts and technical assistance to suppliers of both parts and commodities. Suppliers engaged in these agreements are rewarded with greater volume and contracts for new products. Toyota's manufacturing strategy is counterintuitive to the work of Oliver Williamson, a Nobel laureate, who theorized that decisions regarding whether transactions should take place within a firm or across firms are determined by the need to minimize the costs of transactions. Williamson's transactions cost theory lends itself to asking the following questions:

- First, why sign contracts with firms that lack the required technical capability and help them to develop that capability?
- Second, markets for commodities have low transaction cost hazards and there is little product differentiation, so why offer long-term contracts when one could buy from the cheapest producer?

Helper said that the workshop paper proposes transaction spillovers as an answer. Many economic theories analyze a single transaction at a time, but products generally require multiple intermediate inputs. Spillovers exist across these relationships, increasing the clarity and credibility of contracts with long-term suppliers. A network of tenacious problem-solving suppliers can quickly identify root causes of quality issues and implement solutions. To create tenacious firms, Toyota makes an investment in each firm and benefits by learning from each supplying firm. The supplying firms also learn from each other, as Toyota offers long-term contracts to just a few firms. Due to these network spillovers, Toyota can use a firm-wide strategy for governance instead of a product-level strategy. Clarity in one relationship aids a second relationship even in the case of different parts, which creates an economy of scope for the customer.

Helper stated that in transaction cost analysis, the focus is on individual inputs rather than firm-level relationships. This type of analysis predicts that firms will use spot markets for commodities and enter long-term relationships for high research and development (R&D) aspects of the value chain (Antràs and Chor, 2013; Hart and Moore, 1990; Williamson, 1979). U.S. automotive manufactures appear to follow these predictions; for example, Chrysler sources stampings from Die-Matic, but this relationship doesn't have a long-term guarantee, since stampings are a commodity product with many suppliers. Chrysler sources an innovative crankshaft from Mahle as a sole source provider. Mahle enjoys a long-term relationship and collaboration with Chrysler as its sole producer of an R&D-intensive product. Spillovers do exist in this example, namely as an incentive to reduce switching costs.

CONTRIBUTION

As contributions of the workshop paper, Helper pointed to new insights into the development of supply networks by considering multiple transactions at a time. In this framework, lead firms and suppliers experience economies of scope by developing a clear and credible network. Because of network spillovers, firms and nations have unique and stable strategies for global value chain (GVC) governance, including the number of suppliers per part, degree of vertical integration, and length of relationships with suppliers. The authors applied their framework to U.S. Customs microdata, organized as a panel, on imports from automotive manufactures from 1997 to 2015. They found that transaction-based measures are significant, but organization-based measures generally have much greater explanatory power. This is important as it suggests that response to shock differs by organizational strategy, not just by product; it also suggests that it is necessary to understand not only the patterns of products in trade, but also how those are products are integrating into value chains by lead firms.

CONCEPTUAL FRAMEWORK

Helper pointed out several reasons a firm might only enter purchase agreements in a nonspot/collaborative fashion:

- First, this organizational structure facilitates relational contracts that strengthen the supply network. By creating tight-knit supply networks, firms build clarity and credibility. This allows the lead firm to dictate the specifications of parts more easily, leading to a higher-quality final good. In addition, the supply firm can better withstand exogenous market shocks.
- Second, spillovers exist across functions in the buyer's organization.
- Third, complementarities in adoption of practices and the development of capabilities exist within the network of supplying firms (e.g., just-in-time inventory; Dyer, 1996) and learning to write good contracts (Argyres, 2013).
- Fourth, some buyers may see two products as similar when others in the industry do not; for example, Toyota wants all suppliers to both explore and exploit (Aoki and Wilhelm, 2017).

Helper argued that well-connected supply network spillovers allow for purchasing strategies to interact with other functions in a cohesive manner. Collaborative supply networks incentivize firms to learn from one another, increasing the efficiency of all firms in the network. Supply networks organized at arm's length allow firms to interact only through purchasing, which may reduce average quality across many competing suppliers. Another important dimension is what entity designs the product. Lead firms will have to design predominantly in-house with arm's-length transactions.

Lastly, as Helper described, the career path for purchasing agents differs in the two frameworks. In a collaborative network with strong firm-to-firm connections, purchasing agents can move between a variety of functions, while in an arm's-length agreement, agents should be rotated between commodities to avoid capture. A benefit of the arm's-length network is that agents don't need a deep technical understanding of the product, as they are more reliant on market forces.

To test the propositions presented in the conceptual framework, Helper and Munasib first ran regressions to understand the relationship between product attributes and the style of supply chain governance. Measures for understanding this relationship include product fixed effects, a measure of R&D intensity for a product category, a measure of product differentiation, and a measure of the extent to which an industry is upstream.

Next, the authors tested the theory for firms or nations with organization-wide relational contracts (a collaborative framework). They found that U.S. automotive manufacturers are more likely to engage in spot purchasing relative to Japanese firms. This result is robust to removing variation due to product attributes. The authors also found that Japanese automotive manufacturers are more loyal to their supply network than U.S. firms, as discussed below.

DATA

Helper explained that she and Munasib used microdata from the Longitudinal Firm Trade Transactions Database (LFTTD), which contains transaction-level data by firm, collected by U.S. Customs and Border Protection. All automotive manufacturers assembling cars and trucks in the United States are included in the dataset, using 10-digit harmonized tariff schedule (HTS) categories and providing detailed information about each part (e.g., vulcanized gaskets; washers and other seals; mountings, fittings, and similar articles of base metal). Lastly, the dataset identifies related-party transactions, defined as transactions between a lead firm and supplier in which the lead firm owns at least a 5 percent stake in the supplying firm. This measure differs from the prior literature that typically only accounted for firms that were 100-percent owned by the lead firm.

The sample contains the annual supplier transactions for all automotive manufacturers headquartered in the United States or Japan from 1997 to 2015, and includes 44 countries and 600 products that have been purchased for at least 5 years by a firm headquartered in either the United States or Japan, and 2 years by firms in both groups. If a supplier consists of multiple affiliate firms in one country, then the authors aggregated these affiliates into a single firm. Lastly, the authors excluded purchases from the automotive manufacturer's home country because they do not have data on purchases from U.S. suppliers. For consistency, the authors dropped the same data from Japan.

EMPIRICAL STRATEGY AND MEASURES

Helper described analysis conducted on two broad categories of firms: representative U.S. firms and Japanese-owned firms. Aggregation used original equipment manufacturer–level imports as weights, and the sample was representative of larger market dynamics. There are four outcome variables of interest: number of suppliers per component per year, Herfindahl–Hirschman Index (HHI) of suppliers for a component per year (a horizontal concentration measure), the share of related-party imports of a component per year (a vertical integration measure), and the average number of years supplied by a supplier of a component. At least 25 percent of sales by U.S.-headquartered automotive manufacturers are sourced internationally, while for Japanese firms this number is over 40 percent. In their paper, the authors argue that use of nonspot, or collaborative, purchase agreements from suppliers in the United States is similar to that of foreign suppliers (see Helper et al., 2012).

RESULTS

Helper then introduced the paper's results by noting that there is some support for some predictions of transaction cost theory. As R&D intensity increases to three standard deviations from the mean, the number of U.S. suppliers falls from 8.96 to 7.55, while the number of Japanese suppliers decreases from 3.32 to 3.09. These results are significant, but organization-based theories have more explanatory power. For all levels of R&D intensity, U.S. automotive manufactures have at least seven times more suppliers than Japanese manufacturers. Japanese firms are less sensitive to R&D intensity when choosing the number of suppliers, and even at the highest levels of R&D intensity, U.S. firms use more suppliers than Japanese firms use for commodity parts. Japanese firms are also more loyal, as they are less likely to drop a supplier if an exchange-rate shock occurs.

As Helper described, any of the results discussed so far looked at the country of ownership and therefore a national culture, but the evidence also points to variation on the firm level, indicating that corporate culture is a determining factor. This variation is identified in the number of suppliers used and suggests that U.S. firms have a high variance relative to the mean.

An alternative story to explain the facts and patterns outlined in the work is one of product complexity. Novak and Stern (2008) show that luxury automotive manufacturers with complex parts are more likely to integrate vertically. Novak and Stern find that Japanese firms that produce luxury vehicles engage less in spot purchasing, relative to other luxury automobile manufacturers. This would imply that final product attributes, and not organizational strategy, determine whether a firm chooses to outsource the supply of components or vertically integrate its supply chain. Helper argued that this is unlikely to explain differences between U.S. and Japanese firms, because U.S. firms have more luxury varieties than their Japanese counterparts in the automotive sector.

Lastly, Helper returned to her opening story, the Kojima–Toyota versus Chrysler–Die-Matic comparison. The authors propose that network spillovers due to clear and credible relationships are the reason lead firms give business to a supplier that lacks technical capability. In the United States, a higher discount rate, more concerns about hold-up, and more faith in markets leads to a lower estimated value of a strong network of suppliers. U.S. firms only establish the relational contracts with high-R&D suppliers and maintain an option to exit from these relationships.

SUMMARY

Helper concluded by summarizing the contributions of her work with Munasib, stating that it has contributed to the literature on relational contracts and the importance of network spillovers across nonaffiliated firms in the network. Using U.S. Customs microdata from 1997 to 2015, their results suggest that transaction theory–based measures are significant and have some explanatory power, but organization-based measures have much greater explanatory power. Lead firms and suppliers experience economies of scope when they develop clear and credible relational contracts. This implies that firms and countries have clear and stable strategies for GVC governance.

DISCUSSION

Discussant: Davin Chor (Dartmouth College)

Davin Chor, associate professor of business administration and chair of globalization in the Tuck School of Business at Dartmouth College, stated that the 2020 World Development report highlighted two channels through which trade in GVCs differs from the traditional view of trade: hyperspecialization and durable firm-to-firm relationships. While there is an extensive prior literature on the hyperspecialization of GVCs, less empirical work has been done on firm-to-firm relationships. There is clearly a relational dimension to sourcing decisions for some inputs in the GVC. The data requirements to study the firm-to-firm relationships are very high and can make studying this problem difficult.

The authors overcame this data challenge by using the U.S. Customs LFTTD data, which offer detailed information on the product level. They exploited a feature of the dataset, manufacturer identification (MID), to pin down the identity of each foreign supplier. Caveats to the approach include the focus on U.S.- and Japanese-owned firms only and the exclusion of domestic suppliers.

Chor described the key finding as the large contrast in the input sourcing patterns in U.S.- and Japanese-owned firms. Controlling for product-level variation, Japanese firms have a smaller number of suppliers, higher supplier HHI, higher related-party trade share, and longer supplier longevity. Japanese ownership (organizational strategy) explains more variation in sourcing patterns than alternative product-level characteristics associated with the transaction cost theory.

Chor stated that the work of Helper and Munasib fits in the broader literature, as defined in Bernard and Moxnes (2018) and Antràs and Chor (2021), in four areas:

- First, the literature on firm-to-firm links, which itself has two approaches: the cross-border trade transactions approach (Benguria, 2021; Bernard et al., 2018a; Blum et al., 2010; Eaton et al., 2016b; and Sugita et al., 2014), and the domestic value-added tax data approach (Adão et al., 2020; Alfaro-Urena et al., 2021; Bernard et al., 2019; Dhyne et al., 2021; Demir et al., 2021; Huneeus, 2018).
- Second, the literature on the duration of firm-to-firm links. Work exploiting the MID in U.S. data comes from Kamal and Monarch

(2018), Monarch and Schmidt-Eisenlohr (2017), and Monarch (2021). Using French data, Martin and colleagues (2020) studied this relationship as well. The broad takeaway from past literature is that newly formed relationships are broken quickly; but, conditional on survival, a durable network emerges.
- Third, the literature that discusses relational contracting in specific markets, often in the setting of developing countries. This literature includes the works of Macchiavello and Miquel-Florensa (2018), Macchiavello and Morjaria (2015, 2021), Cajal-Grossi and colleagues (2020), and Brugues (2020).
- Fourth, the literature that studies teardown analysis; for example, Dedrick et al. (2010) and the iPod. Relating to the conference, the work of Surana and colleagues (see Chapter 6) falls in this fourth category.

Chor described this work as an exciting empirical setting, laying in the middle ground between broad cross-sectional analysis and specific industry-level analysis. The automotive industry is a natural industry to study, as it has many components and is deeply immersed in GVCs. The results on the number of suppliers per HTS-10 products is an issue of growing importance given the renewed emphasis on supply chain resilience in the age of COVID-19 and trade frictions. Chor offered insights along three dimensions: sharpening of the empirical work, economies of scope, and alternative interpretations of the Japanese-ownership effect.

The main finding—the number of suppliers—is striking, but Chor pointed out that further clarification on how some results may be driven by the MID construction would enhance the work. As an example, Toyota may have multiple affiliate firms in Mexico listed under the same MID. The authors aggregated all of these firms into a single supplier, which may have created an endogenous result. This would be a larger concern for Japanese-owned automotive manufacturers that source from related parties more often. If the results are robust to this comment, they would be very strong. The unconditioned results on supplier longevity suggest that U.S. firms have longer relationships than their Japanese counterparts; these results flip only when the mean is conditioned on a set of controls.

Chor suggested that the addition of summary statistics on the number of products exported per foreign supplier to U.S. or Japanese firms would be a helpful addition for understanding the role of economies of scope. This is especially true if this measure turns out to be larger for Japanese firms, consistent with anecdotes from Kojima Industries. Further, looking within sourcing networks, are there temporal dimensions to the number of products per supplier, the total value purchased from each supplier, and the quality of each product? A structural model of transactions cost and a fixed cost per supplier, rather than product, may offer some insight into the mechanisms at work. Are there diseconomies of scope if each supplier is responsible for more inputs (Fally and Hillberry, 2018; Kikuchi et al., 2018)?

Chor concluded that the work is a great opportunity to understand network sourcing patterns with a lot of potential for future work, especially in studying U.S. and Japanese GVC resiliency in the wake of COVID-19 supply shocks. Does greater loyalty or a greater number of suppliers help weather these supply disruptions?

During the presentation, one audience member asked about the difference between a U.S. firm, which sees the United States as the home market, and a Japanese firm, which sees the United States as a foreign market. Abdul Munasib, economist at the Bureau of Economic Analysis in the U.S. Department of Commerce and a coauthor of the presented work, responded that there are data privacy concerns when looking at this detailed data. This places artificial constraints on adding outside firm-level data.

8

Keynote Address:
Foreign Direct Investments and Superstar Spillovers: Evidence from Firm-to-Firm Transactions

Paper Authors: Mary Amiti (Federal Reserve Bank of New York and Centre for Economic Policy Research [CEPR]), Cedric Duprez (National Bank of Belgium), Jozef Konings (Nazarbayev University and Katholieke Universiteit Leuven), and John Van Reenen (London School of Economics, Massachusetts Institute of Technology, British Academy, Econometric Society, National Bureau of Economic Research, CEPR, and Society of Labor Economists)

Presenter: John Van Reenen (London School of Economics)
Moderator: Wolfgang Keller (University of Colorado Boulder)

John Van Reenen, Ronald Coase School professor at the London School of Economics; digital fellow in the Initiative for the Digital Economy at the Massachusetts Institute of Technology; and a fellow of the British Academy, Econometric Society, National Bureau of Economic Research, Centre for Economic Policy Research, and Society of Labor Economists, began the keynote presentation by noting that he will be discussing his and his coauthors' new work in this area. Governments often encourage multinational enterprises (MNEs) to locate activities in their countries. MNEs are more productive and can offer higher wages than domestic firms, and their location in a country is often accompanied by new technologies and management practices, making them very attractive targets. MNEs offer spillovers to domestic firms as these new technologies and practices are introduced.

Van Reenen explained that the average quality of management practices of foreign MNEs is higher than that of domestic firms across the income distribution. The trend in domestic management practice roughly follows productivity distribution, with the United States, Japan, and Germany having the highest-quality management, while low-income countries rank lower in management quality. The distribution of average quality of management for affiliates of foreign MNEs is higher for every country in the sample and flatter across the host countries.

The rationale for government policies intended to attract MNEs is deeper than simply targeting high-productivity firms, Van Reenen said, as governments assume that MNE activity will have spillover benefits for domestic firms. Evidence from firm-level case studies (see Iacovone et al., 2015 on Wal-Mex; see also Bloom et al., 2013; Sutton, 2004) indicates that foreign MNEs can work with domestic firms in the supply chain to upgrade technology and practices.

However, as Van Reenen shared, the econometric evidence on MNE spillovers is mixed. Aitken and Harrison (1999) found negative effects from horizontal foreign direct investment (FDI), and Javorcik (2004) found positive evidence from downstream FDI. An issue with the econometric literature is the use of industry-level data on MNE exposure. If spillovers are due to direct relationships, then industry-level analysis is unlikely to identify the effect. Alfaro-Urena and colleagues (2019) exploited variation in firm-to-firm sales data in Costa Rica and found positive effects to total factor productivity (TFP) for firms that enter the supply chain of an MNE, using an event study approach.

Van Reenen and his coauthors wondered if the results from Costa Rica, an emerging economy, would generalize to richer countries. Furthermore, are the spillovers due to activity in the supply chain of foreign MNEs only, or could the spillovers be due to activity in the supply chain of any superstar firm, including firms that are either a large exporter or very large domestic firm? Lastly, if the answer to the prior questions is affirmative, then what is the mechanism to explain the behavior?

Finally, Van Reenen went on to describe the work of Greenstone and colleagues (2010), who studied the productivity gains from FDI in the United States. They compared counties that won a large "million-dollar plant" to those who just lost out. They did this by scraping news stories in *Site Selection* magazine, which contains information about winning and runner-up counties of large-scale infrastructure projects. Greenstone and colleagues (2010) found that incumbent plants in winning counties have higher productivity growth than incumbent plants in runner-up counties. Bloom and colleagues (2019) updated and supplemented the dataset with news coverage from other sources and added management quality as an additional outcome variable. Their report shows a positive overall treatment effect to management quality following entry by the MNE, domestic or foreign, using an event-study framework. The treatment effect is heterogeneous across industries with management spillovers, and larger in industries in which the flow of managerial information is likely to be larger. This suggests that managerial know-how is a significant mechanism for explaining spillovers between MNEs and firms in their supply chain.

FDI AND SUPERSTAR SPILLOVERS:
EVIDENCE FROM FIRM-TO-FIRM TRANSACTIONS

Van Reenen moved to discussing his and his colleagues' new work in this area. Using firm-to-firm panel data consisting of the universe of Belgian firms from 2002 to 2014 and event study analysis, the authors found about 10 percent positive productivity effects after 5 years for firms that enter the supply chain of an MNE. These supplying firms also see an increase in outputs, inputs (labor, capital, and intermediate goods), and exports. The authors also found a positive effect on TFP for entering the supply chain of very large firms, even if they are not globally engaged or heavy exporters. They found no effect from entering the supply chain of nonsuperstar firms.

Van Reenen stated that two mechanisms have been proposed to explain these spillovers: (1) technology transfer, in which treatment effects are larger when the superstar is in an industry that is research and development (R&D) intensive, information technology related, or human capital intensive; and (2) match making, in which the number of buyers increases, particularly within the superstar's network. These results suggest that the benefits of high-productivity firms extend beyond MNEs.

Van Reenan described this work as fitting into five strands of literature:

- MNE spillovers: Aitken and Harrison (1999), Smarzynska Javorcik (2004), Alfaro-Urena et al. (2019), Alvarez and López (2008), Keller and Yeaple (2009), and Keller (2021)
- Higher productivity of MNEs: Bloom et al. (2012), Helpman et al. (2004), Chaney (2014), Antràs and Chor (2013), Eaton et al. (2011), Antràs et al. (2017), Lim (2018), and Dhyne et al. (2021)
- The impact of large firm entry: Greenstone et al. (2010) and Bloom et al. (2019)
- Production networks: Acemoglu et al. (2012, 2017), Liu (2019), Acemoglu and Azar (2020), Atalay et al. (2011), and Iyoha (2021)
- The rise of superstar firms: Furman and Orszag (2018), Autor et al. (2017, 2020), Bajgar et al. (2019), Gutiérrez and Philippon (2019), and De Loecker et al. (2020)

DATA

Van Reenen went on to discuss data sources, sharing that the main data source for this work is the National Bank of Belgium (NBB) business-to-business transaction dataset, which contains the value of sales, all transactions worth more than 250 euros, in all buyer–seller relationships in Belgium based on value-added tax declarations. This is merged with company accounts data from the NBB Central Balance Sheet office. The company-accounts data cover all incorporated firms and include information on sales, labor, capital, and intermediate inputs. FDI information comes from the NBB FDI survey. Data on trade within the European Union were sourced from the Intrastat trade survey, and data on trade outside of the European Union come from customs data. The baseline productivity measure follows the Wooldridge (2009) approach, but this is compared with more recent methods, such as Collard-Wexler and De Loecker (2020), and a simple measure of value added. The authors are working on implementation of the Iyoha (2021) method, but it is not likely there will be significant differences in results.

ECONOMETRIC STRATEGY

Van Reenen said he and his coauthors used an event-study differences-in-differences approach. Superstar firm j is defined in three ways:

- MNE, more than 10 percent foreign-owned and inward FDI;
- exporter, nonwholesaler with more than 10 percent of sales exported; and
- whether it is a large firm, the top 0.1 percent of the sales distribution.

Supplying firm i enters the supply chain at time t. The authors focused on "serious relationships," in which the supplying firm sells more than 10 percent of its total sales to the superstar. The authors also included firm, industry, and year fixed effects.

BASELINE RESULTS

Van Reenen presented two figures displaying his and his coauthors' results, starting with the increase in TFP for firms in the supply chain of an MNE. Five years after entering the supply chain of an MNE, the supplying firm's TFP increases by about 9 percent. The increase in TFP is slightly lagged, taking 2–3 years to significantly impact the supplying firm (see Figure 8-1).

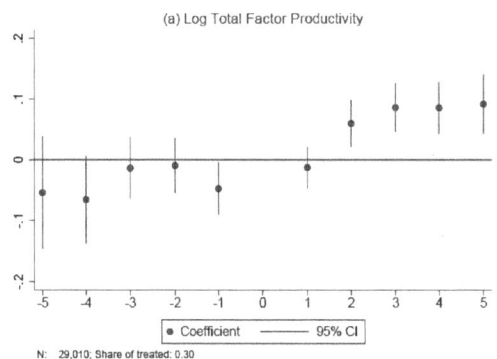

Notes: $t = 1$ first year of treatment; $t = 5$ is all years ≥ 5. Regressions include 4 digit industry by year dummies and firm fixed effects. TFP estimated by Wooldridge (2009) method.

FIGURE 8-1 Selling to multinational enterprises increases TFP by ~9 percent after 5 years: Event study of selling to foreign direct investment firms.
NOTE: CI = confidence interval; TFP = total factor productivity.
SOURCE: Presentation by John Van Reenen.

Next, the authors examined several other outcome variables of interest: total sales, total inputs, total fixed assets, number of buyers, and number of other buyers (those outside the firm's network). For all outcomes listed, there is a significant and positive effect on entering the supply chain of an MNE (see Figure 8-2).

While the analysis above focuses on supplying to MNEs, Van Reenen said that he and his coauthors also studied the effects of supplying to exporting firms. Using the same outcome variables of interest as above, the results are all significant and positive. Lastly, the authors studied the impact of supplying very large firms. Again looking at the same outcome variables as above, there is a significant and positive effect of similar magnitude when a firm enters the supply chain of very large firms.

Van Reenen explained that these results imply that what matters is the relationship between very large firms and firms in their supply chain. However, there is an overlap between global firms and very large firms in a country, so the authors split their analysis into large global firms and large nonglobal firms. The TFP gains for supplying firms are positive and significant for both groups and larger for large nonglobal firms. This indicates that the results are not driven by global engagement, but rather by selling to large firms.

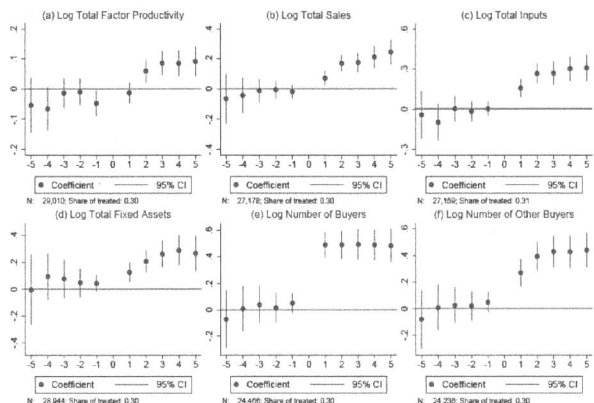

Notes: $t = 1$ first year of treatment; $t = 5$ is all years ≥ 5. Regressions include 4 digit industry by year dummies and firm fixed effects. TFP estimated by Wooldridge (2009) method.

FIGURE 8-2 Selling to multinational enterprises also increases sales, intermediate inputs, capital, and number of buyers: Gains from selling to foreign direct investment firms.
NOTE: CI = confidence interval; TFP = total factor productivity.
SOURCE: Presentation by John Van Reenen.

MECHANISMS

Van Reenen restated that he and his coauthors explored two mechanisms: technology transfer and match making. To understand the technology transfer motive, firms are split into R&D-intensive industries, information and computing industries, and human capital–intensive industries. Regardless of the type of firm (MNE, exporter, or large firm) the results are positive and significant, suggesting that spillovers are larger for high-tech and high-skill industries and that technology transfer is an important mechanism.

The second motive, Van Reenen explained, is match making or the "dating agency" effect. The outcome variables of interest are the number of buyers in the network and the number of buyers outside the network. Across all indicators of firm types, the results are positive and significant. This suggests that the impact on buyers within the superstar's network is strong, although the precise reason for this impact is unclear. It could be that there is a signaling effect when a firm starts selling to a superstar firm, indicating that the supplying firm is of high quality.

SUMMARY

Van Reenen concluded his presentation by stating that firms that enter the supply chain of a superstar firm see positive, nontrivial effects across a range of outcome variables. The main proposed mechanism is the transfer of technology and know-how; match making also plays a role. The authors did not rule out more general spillovers, as this variation is removed with the inclusion of industry by year fixed effects. A main finding is that the firm does not need to be an MNE or globally engaged; the data show that local superstar firms have a similar impact on suppliers. The policy implications of the results are that barriers to firm growth due to misallocation could be costly (see Aghion et al., 2021), supporting government policies that promote the location of MNEs in their country. The next steps in the project include using instrumental variable analysis to understand superstar partnerships and further quantify the results.

DISCUSSION

During the keynote presentation, an audience member asked if there is selection into the MNE/superstar supply chain. In other words, are the results endogenous based on the firm from which the MNE/superstar chooses to purchase? Van Reenen responded that the match-making mechanism is this exact effect. Entering into a relationship with a superstar is beneficial because a firm then gains better access to their network. This is true across all types of firms, and it may be beneficial to investigate whether supplying to MNEs would provide access to foreign networks, in addition to the domestic network. Mary Amati, economist and vice president of the Microeconomic Studies Function at the

Federal Reserve Bank of New York, an associate of the Centre for Economic Research, and a coauthor of the presented work, answered the question by mentioning the breakdown of large nonglobal and large global firms, suggesting that, according to the results, there is no difference between the two.

9

Creation and Diffusion of Knowledge in the Global Firm

Paper Authors: Çağatay Bircan (European Bank for Reconstruction and Development [EBRD]), Beata Javorcik (EBRD, University of Oxford, and Centre for Economic Policy Research), and Stefan Pauly (Sciences Po)

Presenter: Çağatay Bircan (EBRD)
Moderator: Andreas Moxnes (University of Oslo)

The workshop paper by Çağatay Bircan, Beata Javorcik, and Stefan Pauly investigates the diffusion of innovation in multinational firms, which is important for understanding the growth of countries.

Çağatay Bircan, senior research economist at the Office of the Chief Economist at the European Bank for Reconstruction and Development (EBRD), introduced the workshop paper by motivating its importance. The main pillars of modern growth theory are the creation and diffusion of knowledge. Multinational enterprises (MNEs) account for most of private research and development (R&D) expenditures and subsequent innovation activity. The authors used patent data to capture the patterns of creation and diffusion of knowledge in MNEs, focusing primarily on understanding the collaborative process across international borders. In ongoing work, the authors extend their analysis to understanding the role of inventor teams and how inventor characteristics, such as gender, influence the innovation process.

Bircan previewed some of the stylized facts and main findings:

- First, the authors found that knowledge creation is increasingly globally collaborative (defining globally collaborative innovation as patents involving inventors from at least three countries). Further, patents that result from global collaboration are higher in quality, and a large share of MNE patents are invented in a foreign-affiliate firm.
- Second, the authors found that differences in time zones are a major barrier to knowledge diffusion. Barriers due to time zones are larger than direct distance-effect barriers, which are also present and are a

major contributing factor. Time-zone differences also affect patterns of collaboration and citations within an MNE.
- Third, the authors found that diffusion occurs in MNEs through inventor mobility. An overlap in business hours eases barriers to mobility, while distance does not appear to affect mobility. The authors found that, within MNEs, women have less mobility than their male counterparts.

DATA

Bircan explained that he and his coauthors used patent data from two sources: the European Patent Office's (EPO's) PATSTAT database and the U.S. Patent and Trademark Office (USPTO), with coverage from 1980 to 2010. The authors focused on patents granted by the EPO, USPTO, and the Japanese Patent Office (JPO), known as triadic patent families. In the regression analysis, only patents from 2000 to 2010 are included to increase matching quality. The patent data were merged with information about inventor location by geocode and gender. Further, the dataset was merged with Bureau van Dijk's Orbis and Orbis Intellectual Property (IP) databases, allowing the authors to match patent and applicant names using Global Ultimate Owner (GUO). Firms are defined as an MNE if they have affiliates in at least two countries. These firms are the focus of the paper.

STYLIZED FACTS

Bircan presented four stylized facts in detail:

- First, there has been a rise in global collaboration in innovation; however, large variation exists across countries. Japan collaborates the least and its trend appears relatively flat across the sample, with the exception of a small rise around 1995. The United States collaborates at lower rates than Italy, Germany, France, the Netherlands, or the United Kingdom (in order from least to most collaborative in 2010), but these countries generally show a rise in collaboration over the sample, with some decreases. China is the most collaborative country in the sample, peaking at a collaborative share of patents of more than 45 percent, although China's collaboration appears to have flattened in the last third of the sample, and may be falling (see Figure 9-1).
- Second, global collaboration in innovation activity results in higher-quality patents. Across various specifications this result remains positive and significant but varies in magnitude. Patent quality also increases as the number of inventors on the team also increases (see Figure 9-2).

KNOWLEDGE IN THE GLOBAL FIRM

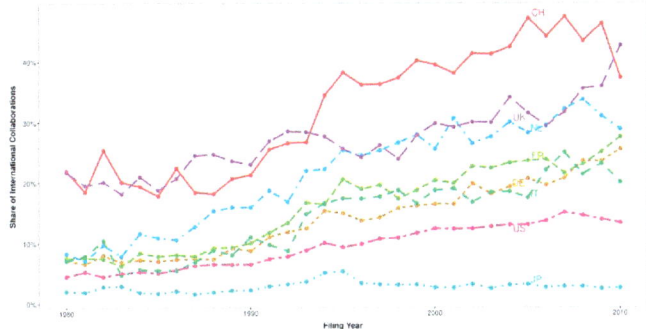

FIGURE 9-1 Cross-border collaboration is on the rise.
SOURCE: Presentation by Çağatay Bircan.

HQ Country	Patent families	HQ inventors (in %)	Affiliate inventors (in %)	HQ-affiliate collaboration (in %)
JP	111606	90.75	4.24	5.01
US	75528	64.41	14.41	21.18
DE	38738	64.56	10.58	24.86
FR	18330	52.17	23.05	24.78
UK	8653	34.59	30.20	35.21
CH	6390	16.07	49.20	34.73
NL	5880	11.36	52.02	36.62
SE	5800	49.72	23.14	27.14
KR	5203	78.47	8.38	13.15
IT	4909	60.87	11.96	27.17

- Similar picture emerges for inventor locations (View inventors)

FIGURE 9-2 Large share of patenting takes place in foreign affiliates.
NOTE: HQ = headquarters.
SOURCE: Presentation by Çağatay Bircan.

- Third, a large share of innovative activity takes place in foreign-affiliate firms. There is a large amount of variation in this share across the sample. European MNEs are innovating especially outside of the headquarter (HQ) country, as are Chinese MNEs. Japanese MNEs have the highest share of solely HQ inventors in the sample.
- Fourth, inventors are more mobile across borders. Similar trends occur with the United Kingdom and China as the highest share of mobile inventors by origin, with Japan having the lowest share. The rest of Europe and the United States appear roughly equal and have a small overall rise in their share of inventors moving by origin. (See Figure 9-3.)

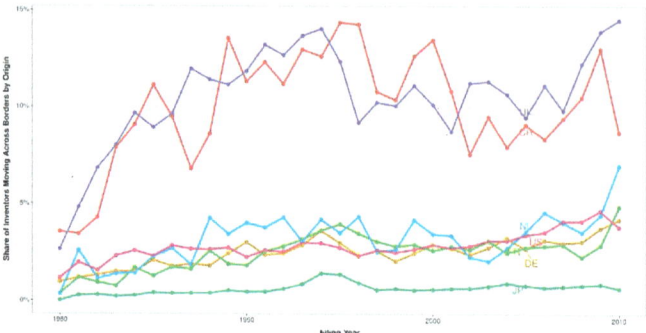

FIGURE 9-3 Inventors have become more mobile across borders.
SOURCE: Presentation by Çağatay Bircan.

EMPIRICAL FINDINGS ON KNOWLEDGE DIFFUSION

Bircan explained that his and his coauthors' empirical findings on knowledge diffusion are split into four categories: global collaboration, citations, inventor mobility, and ongoing work on the role of the inventor's gender. Two empirical approaches are used to understand the first two categories: analysis at the patent-inventor level and analysis at the establishment, or country time-zone, level. The third outcome category uses only the second approach. The first approach double counts patents with multiple inventors, leading to the implementation of the second approach. The sample for the first two regressions contains all patents with at least one inventor outside of the HQ country from 2000 to 2010.

All regressions have the general form

$$Y = \alpha Overlap + \beta Distance + Fixed\ Effects + \epsilon.$$

Overlap is a measure of the business-hour overlap for an inventor in an affiliate-country time zone with respect to the HQ time zone; this measure is specific to the approach used. Distance is a measure of the distance between the inventor country and the HQ country and is specific to the approach used.

Bircan moved to the results in the first category, global collaboration, measured either as an indicator of collaboration in the first approach or as the share of collaborative patents in the second approach. In the first approach, the authors found that both the business-hour overlap and physical distance are significant. The coefficients on business-hour overlap are positive, indicating that an increase in overlap increases cross-border collaboration. Distance has negative coefficients, indicating that the greater the distance between the inventors, the less collaboration occurs among them. To understand the results in a real-world context, Bircan gave an example of a German MNE with affiliate firms in Poland

and Japan. The coefficients overlap implies that the Polish firm would have 33 percent higher probability of collaborating with the HQ, due to the 7-hour overlap in business hours, relative to inventors based in Japan. Likewise, Japanese firms have around a 70 percent lower probability of collaborating with a German HQ than Polish inventors, due to the physical distance between them. These results are robust in the second approach; however, the implied probabilities change when affiliates that do not collaborate are purged from the dataset.

Bircan explained that the second category, citations, studies how foreign affiliates cite patents registered with the MNE HQ. The outcome variable of interest, within-firm citations, indicates whether a patent filed at a foreign affiliate cites at least one patent filed by inventors from the MNE HQ. This means the authors did not examine whether the Polish affiliate cites any German patent, but solely patents from within the MNE. The authors excluded the interfirm patenting activity from the sample, as this may introduce endogeneity concerns between patents. The first approach indicates that business-hour overlap does not play a significant role, but that physical distance contributes a negative and significant coefficient. However, the second approach indicates that business-hour overlap does play a significant role, as does distance.

Third, Bircan and his coauthors examined inventor mobility by studying how inventors move within a firm. The sample is amended from the first two regressions to contain all inventors located outside of the HQ country. The outcome variable of interest is an indicator for whether at least one inventor has moved from an affiliate firm to the MNE HQ, or vice versa. Using only the second regression approach, the authors split their results based on movement from affiliate to HQ, or from HQ to affiliate. In both cases, the business-hour overlap is a very strong predictor of inventor mobility. Distance is not a significant predictor of mobility in either direction.

The authors then considered robustness exercises across the outcome variables of interest.

- First, they included affiliate–HQ-pair fixed effects to explore whether unobserved bilateral variables drive the results, and their results did not meaningfully change.
- Second, to explore whether the results are driven by certain countries, the authors took three approaches:
  – dropping GUOs from the United States or dropping inventors located in the United States,
  – dropping GUOs with multiple home establishments, and
  – clustering at several levels.

Using each of these three approaches within the second broader approach, the authors' results were consistent.

Bircan transitioned to discussing the ongoing work that studies how gender influences the results. Prior literature indicates that women may have different commuting preferences and, for various reasons, may have less mobility. To explore this hypothesis, the authors ran a regression with the main outcome variable as a measure of mobility, and an indicator for gender and a measure of tenure, along with fixed effects. The results indicate that female inventors have much less mobility than male inventors with respect to within-firm cross-border movement. Personal preference and within-firm gender norms are offered as possible explanations for these findings.

To explore the role of within-firm gender norms, Bircan explained, the authors ran a regression with the outcome variable of interest as the share of female inventors employed at the foreign-affiliate firm, along with three explanatory variables: the gender gap between the host and HQ countries using the World Economic Forum inequality index, the gap between girls' and boys' mathematics scores on the PISA (Programme for International Student Assessment) for the host and HQ countries, and the share of female inventors observed at the HQ. The authors found that affiliates in countries with a larger gender gap employ more women, supporting the idea that MNEs can overcome local norms. Further, more female inventors are found in countries in which girls perform better than boys on mathematics exams. Lastly, there is a positive correlation between the share of women employed at the firm HQ and the share employed at the affiliate.

SUMMARY

Bircan and his coauthors presented cross-country evidence on where and how MNEs innovate. They presented several stylized facts that document barriers to the ability of MNEs to diffuse knowledge across borders. They indicated that a major barrier to this diffusion is the overlap, or lack thereof, in business hours between affiliate firms and the HQ, a result that implies that communication costs, monitoring, and repeated interactions are important for innovative collaboration across borders. The authors also presented early-stage work about the role of inventor teams, focusing on heterogeneity of results due to gender and country-level gender norms.

DISCUSSION

Discussant: Heiwai Tang (Hong Kong Business School)

Heiwai Tang, professor of economics at the Hong Kong Business School (on leave from The Johns Hopkins University) and associate director of the Asia Global Institute, began by summarizing the workshop paper's empirical strategy. The main explanatory variables across specifications are business-hour overlap and physical distance. The authors used different measures from the patent data to infer patterns about knowledge creation and diffusion within MNEs.

Tang summarized four main results from the paper:

- First, innovation activity is increasingly global, which leads to higher-quality patents and an increasing share of MNE patents coming from the foreign affiliates.
- Second, time-zone differences and physical distance are barriers to collaborative innovation.
- Third, inventor within-firm mobility is decreasing in time-zone difference and the physical distance between affiliates and HQ.
- Lastly, the mobility results are heterogeneous with respect to the gender of the inventors, where female inventors have less mobility than their male counterparts.

Tang then gave a brief overall assessment of the paper, which, he said, needs little motivation due to the importance of how innovation is created and transferred across MNE networks. The authors presented interesting and important stylized facts about MNE innovative activity. The paper contributes to various literatures on how MNEs divide labor and organize innovative activities within their network, spillovers due to foreign direct investment (FDI), and communication barriers.

Moving to broad comments on the work, Tang pointed out three interesting projects that could come out of the presented work:

- The creation of stylized facts about innovative activity and collaboration by MNEs.
- Further understanding firms' global innovation strategies by going deeper than explanatory variables on distance and business-hours overlap and exploring other country-dependent variables, such as institutions, market size, and knowledge stock.
- The work on inventor mobility and the analysis of heterogeneous outcomes due to gender could be its own work. Currently, it feels like an add-on to an already high-quality paper.

Tang offered five specific comments in addition to the broad comments above:

- Are the results on international collaboration driven by European HQ activities? According to Figure 9-1, Japan plays a nearly meaningless role in global collaboration, and the United States also plays a small role. The results withstand many robustness checks, but the phenomenon itself seems to be European in nature.
- The mechanical results may be driven by the empirical strategy. Patents are only considered if they are filed across all three patent offices, but Japanese MNEs mainly only file with the JPO; as a

result, much of this activity is dropped from the sample. This bias may also be true for U.S. MNEs. Each patent office has different standards for successful applicants, although the USPTO may be considered more global, as the most productive firms choose to patent there in addition to their home office. Tang suggested that the authors may only capture the tip of the iceberg in knowledge diffusion because of this high-productivity bias.

- The authors use an unconventional strategy for identifying affiliate firms. The Orbis database does not identify affiliate firms, but the authors used a creative solution, identifying an establishment as an affiliate when the foreign applicant for a patent is identified. A concern is that affiliate firms that do not file a patent are not recorded, which may alter the interpretation and bias the results.

- The authors see business-hour overlaps as a benefit that eases communication. Tang offered an alternative interpretation, suggesting that these results may be driven by a strategic acquisition motive. An HQ firm may acquire a firm that is innovating in the same space as itself. This may introduce reverse causality concerns, especially for the regressions on citations and similarity, as an affiliate patent that cites an HQ may indicate strategic acquisition. To work around these issues, Tang suggested reintroducing the temporal dimension and identifying an exogenous event, such as the Schengen Agreement in 1995.

- As the authors focus on the communications costs, there should be more insight on heterogeneous effects. For example, different technologies may innovate at different levels—basic, product, or process—and collaboration may be different across these types of innovation. Some technologies require more communication, and these may be more sensitive than others to the business-hour overlap and physical distance. Further, is the innovation groundbreaking or more incremental? Keller and Yeaple (2013) present the idea of "gravity of knowledge," alleging that industries that are R&D intensive have a lower share of routine tasks, which may suggest that they are more reliant on communication. As a final note, Tang suggested that the results are mostly driven by horizontal FDI and wonders if there is a systematic difference between that and vertical FDI.

Tang concluded by stating that the authors presented several new stylized facts that are important for various literatures. The authors may be able to split the work into two or three papers and build a structural model. Beyond that which has already been completed, more work is needed to prove that the mechanism in the business-hour overlap benefit is communication. Lastly, some work is needed on how an MNE will decide where to file patents, in order to better understand countries outside of Europe.

Bircan responded that Tang is correct about the missing foreign affiliates because of affiliates that have no patenting behavior. The authors focused on triadic patents to have a certain level of quality. They can investigate what occurs when they exclude Japan, as it is an outlier in terms of patenting activity. The strategic acquisition motive is interesting, and the authors are currently exploring it. They are compiling a dataset that contains the transfer of patents across firms and are aware of the acquisition motive of MNEs.

Beata Javorcik, chief economist at the European Bank for Reconstruction and Development, professor of economics at the University of Oxford, affiliate of the Centre for Economic Policy Research, and a coauthor of the presented work, responded to Tang's concern that the results were driven by European MNEs, stating that the authors did run their regressions on U.S. MNEs, and their results were robust.

During the presentation, one audience member asked about the use of subsidiary information from the Orbis database. Bircan responded that the authors are not currently using the subsidiary information from Orbis, but they are hoping to do so. Another asked about data on inventor location. An inventor may not be employed by the affiliate but could be employed at arm's length. Bircan responded that, while the authors are not able to certify the absolute location of innovation, they argue that to a first-order approximation, this is a reasonable measure. Bircan noted that the use of the subsidiary information from Orbis may help alleviate some of these concerns. Another audience member mentioned the nonlinearity of the globe. As one moves westward in Europe, one encounters more water; can the authors show their time-zone results on a subregion that is landlocked? Bircan responded that this is a rich area that the authors want to further clarify in their next draft.

10

Firm Selection and Organizational Choice: Complex Patterns of Global Sourcing[1]

Paper Authors: Valerie Smeets (Aarhus School of Business) and Frederic Warzynski (Aarhus School of Business)

Presenter: Frederic Warzynski (Aarhus School of Business)
Moderator: David Chor (Dartmouth College)

Frederic Warzynski, professor of economics at the Aarhus School of Business, discussed the measurement of globalization from the point of view of Denmark using data collected by Statistics Denmark and offshoring surveys coordinated with Eurostat. The workshop paper by Valerie Smeets and Frederic Warzynski, "Firm Selection and Organizational Choice: Complex Patterns of Global Sourcing," studies firm selection and organization choice in the production networks and outsourcings patterns of Danish firms. The presentation, however, drew from a mix of this and two other papers: "Measuring Globalization: A Danish Perspective" (with Peter Bøegh Nielsen, in progress) and "Offshoring, Workforce Composition and Innovation," which is based on "Heterogeneous Globalization: Offshoring and Reorganization" (Bernard et al., 2020).

The work is motivated by the increasing fragmentation in global value chains (GVCs) since the 1990s, the fall of the Berlin Wall, and China's entry into the World Trade Organization, along with the technological shift towards computing and information services. These changes have important implications for labor markets, creating a challenge for policy makers.

Warzynski argued that Denmark is an interesting country for study because it is an early adopter and mover in globalization, and is home to large, multinational enterprises (MNEs) with large market share in their respective industries. These MNEs are relatively open to foreign supply chains in the context of the European Union (Eurostat, 2017). Statistics Denmark creates very detailed

---

[1] The title of Warzynski's presentation was "Measuring Globalization: A Danish Perspective."

and rich datasets, thanks to the openness of Danish MNEs, allowing Danish researchers to answer interesting questions.

LITERATURE REVIEW

The three papers discussed in this presentation relate to an extensive body of literature containing six categories:

- Transaction costs and property rights approach: Coase (1937); Williamson (1975); Grossman and Hart (1986)
- Determinants of organizational form (theory): Antràs (2003, 2014); Antràs and Helpman (2004); Antràs and Chor (2013, 2021)
- Determinants of organizational form (empirical): Tomiura (2007); Du et al. (2009); Kohler and Smolka (2014)
- Measuring offshoring and labor market implications: Feenstra and Hanson (1999); Hummels et al. (2012, 2018); Bernard et al. (2021)
- Measuring GVCs: Olsen and Statistics Denmark (2008); Sturgeon (2013); Sturgeon et al. (2013); Eurostat (2017); Nielsen (2018)
- Offshoring and innovation: Rodríguez-Clare (2010); Arkolakis et al. (2018); Bernard et al. (2021)

THEORETICAL BACKGROUND

Warzynski discussed the theoretical background for his presentation by describing a simplified theoretical framework based on Antràs and Helpman (2004), which applies the theory of firm and property rights in a global context. In this model, firms differ in two dimensions, which depend on the nature of the firm's activity: productivity and headquarter (HQ) intensity of a needed input. When HQ intensity is low, adding little value at the HQ, control is not important and the efficient strategy for the firm is to outsource activity to countries with comparative advantage in the required factor of production. High-productivity firms will outsource to the global South and low-productivity firms will outsource to the global North. The least productive firms will then exit the market.

Conversely, Warzynski said, when HQ intensity is high, or there is a large amount of value added at the HQ, then control and location are important decisions. The most productive firms will produce in the global South, with foreign direct investment (FDI), and somewhat less productive firms will outsource to the global South. The firms below these somewhat less productive firms in the distribution will source components in the global North, either through integration (for the more productive firms) or outsourcing (for the less productive firms). Again, the least productive firms in the distribution will exit the market. Warzynski and his coauthors tested these theories using the Danish data described below.

DATA

Warzynski went on to explain that data were collected from three waves of mandatory offshoring surveys, each containing more than 4,000 firms, from 2001 to 2006 (reported in 2007), 2009 to 2011 (reported in 2012), and 2014 to 2016 (reported in 2016). All nonfinancial firms in Denmark that have more than 50 employees were required to take part, and the surveys had a compliance rate of more than 95 percent. In addition, there is a subsample of firms with between 20 and 49 employees.

Warzynski said that firms were asked about the type of business activity they outsourced. Firm activity was categorized into core activity and support activity. Core activity describes firm functions accounting for their main source of revenue. Support activity is divided into subfunctions: distribution and logistics; marketing, sales, and after-sales services (like support centers); information and computing technology services; administrative and management; engineering work and other technical services; research and development (R&D); facility management (only asked in the first round); and others. Firms were also asked about the organizational form of their outsourcing activity broken down by intra- and interfirm. The destination choice is broken down by "old European Union" (western Europe), "new European Union" (Central and Eastern Europe), China, India, other Asian countries and Oceania, and others, such as other European countries outside of the European Union, the United States, or Canada.

In the work presented, the focus was on the firm's core activity only; Warzynski argued that this has the largest implications for the domestic labor market and represents the most important decision for the firm organizational structure. The findings can be extended to include firms' support activities as well. The focus on business functions is discussed in depth in Olsen and Statistics Denmark (2008), Sturgeon (2013), Eurostat (2017), and Nielsen (2018), and is inspired by Porter (1985). The general idea is that firms separate and choose their activity in an efficient manner, and some activity is more likely to be offshored. Firms can also acquire a competitive advantage by choosing the correct level of activities of each form. Breaking down firm activity this way is a "new tool in the statistical toolbox" (see Nielsen, 2018) and can be validated by use of occupational groups in datasets linked by employer–employee relationships, as in Bernard et al. (2017) and Bernard et al. (2021).

SUMMARY STATISTICS

Warzynski first discussed the offshoring activity for both core and support activities. In the first wave (2001–2006), a large proportion of firms (more than 40 percent) purchased from domestic firms, driven by the category of facility management (e.g., cleaning, food). Facility management was not included in the other survey waves. Also in the first wave, approximately 6 percent of firms sourced only internationally and about 11.5 percent sourced both domestically and internationally. About 41 percent of firms did not outsource any activity in the first wave. In the next two survey waves (2009–2011 and 2014–2016), domestic sourcing dropped significantly to about 19 percent in the second wave and 18 percent in the third. The number of firms that did not outsource increased in the second and third waves to 64 percent and 72 percent, respectively. International sourcing fell from 11.5 percent in the second wave to about 5.5 percent in the third. Firms that sourced from both domestic and international firms remained relatively flat between 2001 and 2016, although there was an increase in the second wave that disappeared by 2016.

Next, Warzynski looked at the core activities only, for which the patterns remained relatively similar. There was a decrease in international sourcing, particularly between the second and third waves (reported in 2012 and 2016), and an increase in domestic sourcing across all waves. There was a small increase, from 1 percent to 2 percent, in sourcing activity both internationally and domestically across the full sample. International sourcing occurred increasingly at the intrafirm level. Interfirm sourcing was relatively flat across the sample with a small dip reported in 2012. International sourcing both within and outside of the firm was flat in the first two waves before decreasing by half in the last wave. International sourcing to the global North was relatively flat with a drop reported in 2012, while international sourcing to the global South is increasing over the three surveys. Sourcing to both the global North and global South is decreasing across the three surveys. The most common destinations for outsourcing, in order of proportion as reported in 2016, were new European Union, China (a decreasing choice over the sample), other Asian countries and Oceania, other European countries, India, and the United States and Canada.

Warzynski summarized the trends as a clear slowing of outsourcing internationally and relatively flat domestic sourcing, which is complicated because of the dropping of the facility management activity. The international sourcing that remains is less concentrated in Asia and even less concentrated in the European Union. The global South remains the majority destination for international sourcing, and fewer firms outsource to both the South and North. International sourcing within a firm is more likely than outside of the firm, and fewer firms outsource to both areas.

ORGANIZATIONAL CHOICE

Warzynski went on to explain that, following the stylized facts of the data, firms are divided into several categories based on their organizational structure, focusing on their core activity only. Firms are split into five categories by whether they outsource within or outside of the MNE, or both: no outsourcing, domestic-only outsourcing, international sourcing outside group, international sourcing within group, and international sourcing both within and outside group. The second categorical variable adds the dimension of location of outsourcing, divided into eight subcategories: no outsourcing, domestic-only outsourcing, international sourcing outside group in the North, international sourcing outside group in the South, international sourcing within group in the North, international sourcing within group in the South, international sourcing within and outside group in the North, and international sourcing with and outside group in the South.

ADDITIONAL SURVEY QUESTIONS

Warzynski stated that the second two waves of the survey asked firms to define their core activity, in order to determine whether the firm is primarily engaged in manufacturing, trade, or services. Firms primarily engaged in manufacturing were further subdivided into three categories: regular manufacturer, subcontractor, or a factory-less goods producer. Firms primarily engaged in trade define themselves as regular trade or a trade firm involved in manufacturing. Firms engaged primarily in service are divided by regular service and services involved in production. There is large heterogeneity within sectors, implying that firms engage in a mix of the three broad categories.

Moving to employment by business function, Warzynski said the second two surveys asked firms about the distribution of employees by function. This is useful for understanding the activity performed at the firm and how the firm allocates resources over time. This reporting was validated with occupation codes in a merged dataset (see Bernard et al., 2017, 2021).

Warzynski mentioned that the second two waves of the survey investigated the flow of activity back to Denmark. Firms were asked about their backsourcing activity, providing systematic evidence about the global sourcing decisions of Danish firms. Responding firms indicated that most of them were not involved, and the ones that were involved did not indicate that many jobs were created.

TESTABLE PREDICTIONS AND EMPIRICAL APPROACH

Warzynski explained that he and his coauthors used multinomial logit regressions with the two measures of organizational structure as the outcome variables of interest; the independent variable is the probability of a given firm choosing a particular form of organization, with explanatory variables of value added per worker, capital per worker, size of firm, a manufacturing indicator, and second-wave indicator.

Warzynski and his coauthors found that international sourcing outside of the group is negatively associated with capital intensity, a proxy for HQ intensity. Further, this sourcing is positively associated with value added per worker, a proxy for productivity. For international sourcing within group, and both within and outside group, value added per worker is positive and significant, while the proxy for HQ intensity is not statistically significant.

The results, Warzynski shared, show a link between decisions about organizational choice based on firm productivity and HQ intensity. HQ intensity is negatively related to outsourcing and positively related to FDI and complex sourcing. Firms on the right tail of the productivity distribution are more likely to source in the global South and more likely to organize through FDI, conditional on HQ intensity. The results are robust when looking at core and support activities, not core alone, and using the stock of activities rather than the flow. These results validate the predictions of the work of Antràs and Helpman (2004) and the extension to bisourcing by Du and colleagues (2009).

Warzynski mentioned that labor market implications from these surveys can be found in Bernard et al. (2021), which reports that companies that outsource activity to the new European Union or China become more innovative and upgrade their workforce by hiring more high-tech, high-wage employees. That work merged the first wave of the study with R&D surveys, employer–employee data, a production survey, and international trade statistics from customs data. Nielsen (2018) studied large industrial firms in the second two waves of the survey, finding a large decrease in the number of unskilled workers at outsourcing firms and an increase in the number of skilled workers between 2009 and 2016.

FUTURE DATA COLLECTION

Warzynski indicated that the fourth wave of the survey would launch in August 2021 and contain similar questions, in addition to asking firms how the COVID-19 pandemic impacted GVCs. He and his colleagues will continue to integrate additional data sources.

SUMMARY

In conclusion, Warzynski stated that his and his coauthors' work shows evidence of complex outsourcing patterns involving intra- and interfirm sourcing in the global North and South. Outsourcing decisions appear to be driven by two measures of firm heterogeneity: productivity and capital intensity. These results validate the theoretical predictions by Antràs and Helpman (2004). In future work, the authors will integrate more data sources and seek a better understanding of sourcing decisions in line with the make-or-buy literature.

DISCUSSION

Discussant: Catherine Thomas (London School of Economics)

Catherine Thomas, associate professor of managerial economics and strategy in the Department of Management at the London School of Economics, director of the Centre for Economics Performance International Trade Programme, and a fellow of the Centre for Economic Policy Research, discussed only the submitted paper, "Firm Selection and Organization Choice: Complex Patterns of Global Sourcing."

Thomas began by stating that this paper is based on firm-level survey data from Danish firms. In the results reported in 2007, there were more than

4,000 firms divided in two non–mutually exclusive dimensions: intra- or interfirm boundaries, and domestic or international location. According to Thomas, understanding the variation in firm organization choices will reveal information on the firms' relative efficiency, cost-reduction, or allocative efficiency, and eventually enable analysis of the welfare implications of the firms' choices.

As Thomas reviewed, the authors connected the variation in the data to theory developed in the past literature, most notably the Antràs and Helpman (2004) model on how to mitigate the impact of incomplete contracts. The use of theory to understand the data limits the possible sources of variation that may exist outside of the formal framework. The model contains a production function with inputs from two parties with variable fixed costs and heterogeneous firms, as a result of variation in productivity and relative input factor intensity. In the formal model, productivity is measured firm-wide, which is difficult in the context of an MNE. This theory produces predictions about the sorting of firms into four mutually exclusive organization choices, depending on model parameters.

Thomas pointed out the benefit of this rich firm-level dataset as reducing required assumptions for analyzing the predictions of the model. The authors can explore what must be true for the predictions of Antràs and Helpman (2004) to be true. If these requirements are plausible, then one could begin to argue that the location decisions are due to incomplete contracts, as in the formal model.

Thomas then stated that the results indicate that firms select intrafirm sourcing when they have high value added per worker. This is consistent with the formal model with high HQ intensity and moderately high relative fixed costs of integration.

On the other hand, Thomas continued, the results indicate that this association only exists for firms sourcing in the global South, not in the North. She raised an additional concern that the productivity measure is not firm-wide as in the formal model, but rather is for the domestic HQ only. There may be alternative explanations for the observed correlations. More productive firms have more scale capacity to incur the fixed costs of fragmentation in their GVC. The firm then alters the tasks that they complete in-house, and firm activity that adds more value can be done at the firm HQ. Activities with less value added can be outsourced. Then efficiency gains arise from comparative advantage and not from the need to mitigate underinvestment.

In order to validate the results, Thomas stated that the theoretical model makes unique predictions that the authors can explore. One avenue may be to split the data by productivity and capital intensity, in order to understand how heterogeneity impacts their results. The model predicts different relationships under different model parameters. The results would be convincing if these relationships existed for firms that match those parameters. Further statistics on bargaining power and contracting between firms would be necessary to truly fit the proposed theoretical framework.

Thomas said the data indicate that some firms are entering extremely complex organizational structures with combinations of integration and outsourcing at home and abroad. There are several possibilities for the complementarity between strategies. Du and colleagues (2009) suggest that using multiple sourcing of the same input is a strategic decision by a firm to improve its bargaining power relative to its suppliers.

Finally, Thomas concluded with these questions: Do economies of scope influence sourcing decisions of large, productive firms? Are there lower fixed costs to outsourcing an input if a firm already produces another input in-house in that location? Firms may "learn by doing," thereby allowing them to easily operate an additional affiliate if they already have one. There may also be benefits to coordination among affiliates that are integrated. Characterizing "global value networks" may motivate new models of selection.

Warzynski began by addressing the question about the measure of firm productivity. A global measure of MNE productivity is not yet defined and will hopefully be developed. The authors have attempted to validate the model predictions, but the lack of complete validation may generate new theories on these mechanisms.

Warzynski also said that he and his coauthors are exploring whether scale and changing tariffs might provide comparative advantage related to the dynamic decisions of firms. The interaction between capital and productivity is an interesting angle, which is something the authors will continue to explore.

Additionally, Warzynski said that the authors want to further explore complementarities of strategies that may help extend the Antràs and Helpman (2004) model.

During the presentation, one audience member asked how the firm-level surveys may benefit input-output tables and if Statistics Denmark is taking any steps to make this improvement. Warzynski responded that he is mostly interested in the micro measurement picture, or the level of measurement of the presented work. Statistics Denmark is working with the Organisation for Economic Co-operation and Development and Eurostat, among others, to aid the development of the macro picture in input-output tables. Another asked a question about the decrease in outsourcing, particularly to China. Warzynski responded that there is less outsourcing to the South in general. He added a caveat that their data are in terms of flows and not levels. Another question arose about the bisourcing question Thomas raised regarding the Du et al. (2009) work, which the Warzynski and Smeets (submitted) paper attempts to explore. They extend the past literature to include location and not solely the make-or-buy distinction.

11

Are Customs Records Consistent Across Countries?

Paper Authors: C. J. Krizan (U.S. Department of Labor), James Tybout (The Pennsylvania State University, National Bureau of Economic Research [NBER], and CESifo), Zi Wang (Shanghai University of Finance and Economics), and Yingyan Zhao (George Washington University)

Presenter: James Tybout (The Pennsylvania State University, NBER, and CESifo)
Moderator: Andreas Moxnes (University of Oslo)

The workshop paper by C. J. Krizan, James Tybout, Zi Wang, and Yingyan Zhao studies firm-to-firm datasets to understand their accuracy for formal hypothesis testing. Accurate data sources are vital for making informed policy choices.

James Tybout, professor of economics at The Pennsylvania State University and member of the National Bureau of Economic Research and CESifo, introduced the workshop paper by motivating the importance of accuracy in measurement of trade statistics and customs data. Researchers have used customs records to understand international trade from the perspective of the firm. This includes understanding a firm's market entry and exit decisions and market share development. These datasets are also used to study the patterns of diffusion of technology across global value chains (GVCs) and multinational enterprise (MNE) networks. The literature on these topics relies on micropatterns in the data to understand these complex issues. Some examples of these studied patterns are firm-to-firm connections and the duration of relationships in GVCs; the frequencies and patterns in shipments, product classifications, and industries; and the value of traded goods.

According to Tybout, the work presented contributes to a literature on the accuracy and reliability of customs data and improves researchers' understanding about the dimensions under which the data perform the best. Tybout and his coauthors compared the export records of shipments leaving Colombia with the import records collected when the shipments enter the United States. The analysis was done at the aggregate, industry, firm, and transaction levels. He highlighted that, as expected, the results are more accurate at the aggregate and industry levels versus the firm or transaction levels. He and his coauthors offer alternative explanations for the observed discrepancies, possible implications for research in the field, and some suggestions for improvements in tracking records.

While the paper studies only the Columbia–United States customs data discrepancies, these discrepancies are also true for trade across other U.S. trade partners. According to a 1996 U.S. Census report, the discrepancy between U.S. and Australian trade is around 4–7 percent. Orsini and dos Santos (2015) show that the gap for Brazil and the United States is between 11 percent and 17 percent, while the 2012 U.S. Census estimates the gap between the United States and China to be somewhere between 22 percent and 48 percent. Kellenberg and Levinson (2019) find large variation across 126 countries.

LITERATURE REVIEW

Tybout started by explaining how this work fits in the related literature:

- Record matching: The literature on record matching shows that it is a powerful tool (see Christen [2012] for an overview).
- The Longitudinal Firm Trade Transactions Database (LFTTD): This database, developed in Bernard et al. (2009) and further curated in Barresse et al. (2017), Kamal and Monarch (2018), and Kamal and Ouyang (2020), contains the U.S. Customs data.
- Trade reconciliation studies: This literature, which includes U.S. Census (1996, 2012), Orsini and dos Santos (2015), Fisman and Wei (2004), Mishra et al. (2008), Stoyanov (2012), Ferrantino et al. (2012), Javorcik and Narciso (2017), and Kellenberg and Levinson (2019), seeks to understand the variation and discrepancies in trade statistics; however, none perform this analysis at the transaction level.
- Matched customs records studies: A very large body of literature uses matched customs records—Eaton et al. (2014), Blum et al. (2010, 2018), Bernard and Dhingra (2015), Dragusanu (2014), Kamal and Sundaram (2014), Sugita et al. (2019), Monarch and Schmidt-Eisenlohr (2017), Bernard et al. (2018b, 2018c), Carballo et al. (2018), Monarch (2019), and Helper and Munasib (see Chapter 7 of this volume).

STYLIZED FACTS AND TRENDS

Tybout displayed the reported exports from Colombia and reported imports from the United States from 2007 to 2013, as well as the percentage difference between the two measurements (see Figure 11-1). Prior to 2011, Colombia reported higher trade than the United States, reaching about 12 percent in 2009. Following 2011, the United States increasingly reported more trade than

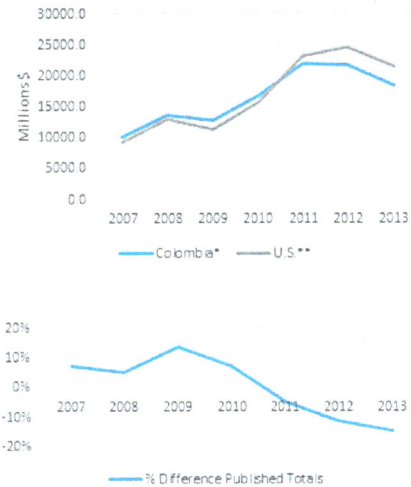

FIGURE 11-1 Official aggregates: Columbian exports (free on board) to the United States.
SOURCE: Presentation by James Tybout.

Columbia, reaching about 15 percent in 2013. Three industries—ceramics, nonknitted apparel, and knitted apparel—account for the majority of Colombia–U.S. trade. The industry-level analysis shows similar discrepancies between Columbian exports and U.S. imports. This suggests that the measurement issue is on a larger scale than a single industry.

According to Tybout, U.S. Census experts suggest that it is common for U.S. importers to split transactions for administrative purposes, which may cause concerns of double counting in the U.S. data. The U.S. LFTTD data suggest that there are about 8 percent more transactions reported by the United States than by Colombia. At the same time, the LFTTD reports 12 percent higher value for the United States. Because there are reporting discrepancies for both the number of transactions and the size, or value, of those transactions, it is unlikely that the discrepancy is just a matter of double counting by one country.

Another potential reason for record-matching discrepancies may be entrepot trade, Tybout explained, a situation in which goods from one country (e.g., Columbia) are first shipped to another country (e.g., Panama). The second country is then recorded as the destination, although the goods are, in reality, being routed to their final destination (e.g., the United States). If there are intermediate stops in the shipment of goods, exporters may not know the destination country and importers may not know the country of origin. Disentangling this effect is difficult. This problem has been explored in Ganapati et al. (2021), which reports that 80 percent of trade is indirect. Tybout and his coauthors checked for this discrepancy by comparing the reported country of origin and the shipping country.

As Tybout explained, while he and his coauthors found evidence that entrepot trade is increasing and may contribute to the observed discrepancies, entrepot trade did not explain the growing gap in the aggregate trade series. The value of trade by shipping country and country of origin track very closely over the period in question.

METHODOLOGY

Tybout then described the work in the paper, which focuses on the types of misreporting and not the underlying incentives. He and his coauthors explored whether customs records accurately characterize firm-to-firm trade patterns, and if the records match the correct firms in a relationship and their accuracy on the transaction level. They wanted to know whether particular firms or industries are driving the trends in misreporting. Finally, their paper explores the implications of the misreporting patterns for using this data for other questions.

Tybout went on to explain that the two methods for matching importers and exporters in the customs data include an importer identifier (the name and address of the firm) and an exporter identifier (the manufacturer ID [MID]—a string of name and address—for the U.S. data, and NIT—a form of tax ID—for the Colombian data). The authors used the importer identifier since there is less data loss in the matching process. The exporter identifier (MID and a pseudo-MID), is examined briefly but its use was not the preferred approach. Once firms are matched, the authors move to linking the transactions between the pairs.

The authors used a three-step process to merge the LFTTD and Colombian data:

- First, a standard data-cleaning process, developed for the creation of the LFTTD data by the U.S. Census Bureau, was conducted.
- Second, the first matching stage links the two datasets on the firm level using only the names and addresses. In order to reduce computation time, the information on transactions was not used. The result is a list of importer–exporter pairs in which both agents listed the other as a partner in a period, with some noise.
- Third, firm pairs were matched on their transactions by using the industry, product, value, and date as unique identifiers. The resulting dataset contains a list of transactions for each trading pair in a roughly consistent manner.

RESULTS

Tybout continued with a discussion of the results report in this workshop paper. The LFTTD database identifies 9,400 U.S. importers, but only 2,500 of the importing firms are matched between the two databases. However, the 2,500 firms account for about 85 percent of the value of exports and about 73 percent of all transactions, providing robust coverage of a large amount of the trade. Looking at the data from the Colombian side, the 2,500 firms account for 95 percent of the transactions and about 97 percent of the value of trade. These numbers are the aggregate results. At the transaction level, only about 27 percent are matched in the LFTTD data and about 28 percent in the Colombian data. In terms of trade value, about 48 percent in the LFTTD data and about 52 percent in the Colombian data are matched. This means that the matched transactions of the matched firms are less than 30 percent of the total transactions and about 50 percent of the value of exports to the United States.

Over time, Tybout said, the value match is consistently higher than the transaction match. The Great Trade Collapse, which forced smaller exporters to exit, increased the match rate for both value and transactions because smaller exporters were less likely to match in the data.

Match rates also vary with industry type. The authors focused on the three largest categories of Columbian exports to the United States: ceramic products (HS 69),[1] nonknitted apparel (HS 62), and knitted apparel (HS 61). The match rate is highest for the ceramics industry, though it is still below 50 percent. Tybout offered a possible explanation, stating that these industries, especially the apparel industries, have many small firms with high turnover, so they are less likely to match. These trends are confirmed on the value level, but match rates are slightly higher across the board.

Matching also depends on whether firms are related or at arm's length, explained Tybout. Using the LFTTD data, related parties are more likely to match on transactions. Related parties are less likely to match on the volume dimension. This pattern is reminiscent of Bernard and colleagues (2006), who found a similar pattern when studying transfer pricing.

Tybout offered two other, disconnected trends:

- Larger-value transactions are more likely to match than smaller-value shipments.
- The wholesale retailers match better than nonwholesale retailers when matching on transactions. The two groups are similar when matching only on value.

IMPLICATIONS OF RESULTS

Tybout stated that these discrepancies mean that research using firm-to-firm network GVCs could find imaginary links in networks and may be mismeasuring the duration of relationships. Researchers may also be mismeasuring entry, search, and learning costs associated with analysis of

---

[1] HS refers to the harmonized system codes, a standardized numerical method of classifying traded products used by customs authorities around the world.

exporters, in addition to mismeasuring interaction effects in analysis of technology diffusion.

Government statistical agencies should be concerned about these discrepancies in the data because these data are used to enforce commercial policy. If the data are poor, bad actors abroad will not be easily identified. There is a clear benefit to increasing the accuracy of firm-to-firm and transaction-level data.

BETTER IDENTIFIERS

Tybout presented some possible alternative identifiers for transaction-level data. Shipment invoice numbers are issued by the seller to the buyer, and these invoices list characteristics such as product and price. This is the main documentation of a sale between a firm pair. U.S. import declarations require an importer to verify and attest to records accuracy, but there may not be a similar requirement of information on the export declaration in foreign countries.

Another proposed identifier Tybout discussed is the bills of lading (BOL) number, which is issued by shipment carriers. BOLs establish a receipt of the product and contain evidence of title of ownership of the products, although these records refer to multiple invoices and identify a container rather than a shipment, and may include multiple containers. In the United States, the BOL can be used to identify the importer, but the foreign customs records may or may not identify an importer.

Tybout went on to say that the two proposed identifiers require global coordination between governments, which may not be feasible. An alternate identifier from the private sector would be Dun and Bradstreet numbers, or other similar firm-level identifiers. While there are major issues with these identifiers, collecting data on subsamples could still be valuable and provide a sample of true and false matches to compare with other datasets. This would open the possibility of matching algorithms based on a training sample.

DISCUSSION

Discussant: Jeronimo Carballo (University of Colorado Boulder)

Jeronimo Carballo, assistant professor of economics at the University of Colorado Boulder, began by motivating why the work is important to research in international trade and GVCs, which both play a defining role in the world economy. This literature depends on the use of these firm-to-firm databases, so accurate measurement is very important. This paper is the first to document the accuracy of firm-to-firm transaction data. The authors explore the consistency of data along dimensions, taking their study beyond trade value by examining transaction and firm links. This difficult task has much value. The main takeaway is that firm-to-firm datasets using names and addresses to match must be used with caution, but datasets that match on tax ID may still be valuable.

Carballo first noted that these discrepancies exist for other countries and data sources beyond the U.S.–Colombian relationship. For example, Italian figures are 10 percent higher in Eurostat than the figures published by the Organisation for Economic Co-operation and Development, and Canada reports $20 billion more exports to the United States than the United States reports as imports. The discrepancy is apparent in the data on trade for many countries and the United States, both positive and negative.

Carballo wondered if there is anything unique about the United States relative to the rest of the world. U.S. firms are identified using employer identification number (EIN), a tax identification number. A downside to this identifier is that firms can, and often do, have several EINs, making matching difficult. The Census Bureau has firm-level EINs only for income and payroll tax filings, while a firm may use a separate EIN or completely different identifier, such as a Social Security number or some foreign identifier, for trade shipments. He wondered how many of these discrepancies can be explained by firms having multiple EINs, and suggested that looking into differences between single-unit and multiunit firms may offer some insight into this concern. MNEs can have complex ownership structures, exacerbating the concerns about multiple EINs and obscuring which affiliate is reported in each database. To explore this further, the authors may be able to use Compustat-FirmID to understand if this problem is a larger concern for MNEs.

The LFTTD database contains information on two firms in each import: the ultimate consignee (the firm that takes final control of the good) and the importer. For about 20 percent of transactions in the United States, the importer is different from the ultimate consignee, and understanding what entity is reported in the foreign database is important, as this may offer an area from which missing links can be recovered.

Carballo also suggested that the authors explore the port of entry to test whether entrepot trade is the source of the discrepancy. The West Coast of the United States may be more sensitive to this effect than the East Coast. He wondered if the authors have data at the port level that would allow this style of analysis, namely if the Colombian data have the same accuracy as the LFTTD. In a similar vein, he wondered if the results vary by mode of transportation as transshipment is more common for maritime trade.

Carballo also wondered if discrepancies vary based on the mode of transportation and whether relaxing the temporal dimension would increase matching rates. Maritime trade may potentially take several days, while air freight may occur in one. Further, shipments may wait at a warehouse or port, and this would cause delays that may impact the time linages.

Finally, Carballo offered several minor comments and questions:

- Why did the authors drop the agriculture and mineral products?
- The cutoff is different for exports and imports in the United States; are the patterns similar for each?

- What are the consequences of matching at the transaction level using a many-to-many approach?

Tybout responded by saying that he and his coauthors dropped agriculture and mineral trade because the numbers failed to make sense. Trade in oil will sometimes appear as an export from the United States to the United States, and there is a central coffee exporting firm in Colombia, so the unique identifier is lost. He also responded that they tested the temporal dimension, but it did not appear to be very important. In the future, they will explore mode of transportation and ports of entry.

C. J. Krizan, director of data analytics at the U.S. Department of Labor and a coauthor of the presented work, commented that he was at the Census Bureau for 15 years and was very aware of the concerns with the EIN as an identifier. The Census Bureau does try to build a unique firm identifier that collects all the EINs a firm may have. In this project, the authors attempted to match on this unique firm identifier, but this did not have a large impact on the results.

One audience member asked about differences in the definition of a transaction, especially in the textile industry. Tybout responded that the authors aggregated all transactions in a firm pair, and they still did not recover a high-quality match. Krizan added that splitting shipments is common for U.S. firms; this may impact the match rates.

12

Capital Flows in Global Value Chains

Paper Author/Presenter: Xiang Ding (Georgetown University)

Moderator: Nadim Ahmad (Organisation for Economic Co-operation and Development)

The workshop paper by Xiang Ding studies capital flows in the context of global value chains (GVCs). This is an important consideration to add to structural models to develop a more nuanced and fuller perspective of international trade.

Xiang Ding, assistant professor of economics at the Walsh School of Foreign Service at Georgetown University, introduced the workshop paper by motivating the importance of understanding the flow of capital in GVCs. He provided a stylized example about the semiconductor GVC to motivate. Manufacturing of semiconductors takes place in Taiwan with silicon coming from China, capital equipment from Japan, and research and development (R&D) from the United States. This means that both tangible and intangible capital are durable intermediate inputs in the production process and can be traded in a similar fashion to nondurable inputs, such as silicon. In the context of trade liberalization, while it is understood how liberalization between China and Taiwan would reduce the price of semiconductors, understanding how liberalization between Japan and Taiwan impacts prices is just as important.

Ding explained two major challenges in accounting for capital in GVCs: measurement and dynamics. Traditional national accounts tables do not account for the supply-use flow of capital goods. In the example outlined above, an input-output table would be split into three categories: intermediate inputs (China), gross output (semiconductors), and value added (United States, Japan, and Taiwan). In the standard framework the value added term does not differentiate rents due to capital, R&D, or value added due to labor. The second challenge, dynamics, is summarized as "investment today" and is dependent on future expectations. The existing literature treats capital as a primary factor of production and investment as final use in the national accounts.

CONTRIBUTION

Ding stated that his paper contributes to the past literature by proposing a unified framework that quantifies GVCs by treating capital services and intermediate goods alike. The framework is based on an open economy in a dynamic general-equilibrium model that focuses on the long-run steady-state outcome. The model uses the same underlying quantitative framework as the existing static models of trade employed in the field, with a slight alteration to the calibration of the production function parameters.

Ding explained that this framework motivates the measurement of the capital coefficients through understanding the flow of investment by linking the rental payments to capital to the value of new investment. His empirical work uses data from the U.S. Bureau of Economic Analysis (BEA) Capital Flow Tables and the standard World Input-Output Tables (WIOT).

Ding stated that his model predicts new production links between the supply and use of capital. The gains from trade liberalization are more than doubled in the new framework and distributed more heavily in capital-intensive industries. In this instance, capital intensity is for use in both production and supply of final products, or the forward and backward linkages discussed in the workshop paper presented by Shang-Jin Wei (Wang et al., Chapter 3).

LITERATURE REVIEW

According to Ding, this paper fits into or expands the following related literature:

- quantitative models calibrated using WIOTs, which do not consider capital as an intermediate input (Antràs and Chor, 2021; Carvalho and Tahbaz-Salehi, 2019; Costinot and Rodriguez-Clare, 2014);
- existing literature on trade in capital, which focuses on either physical capital or knowledge capital, but not both (Eaton and Kortum, 2001; Keller and Yeaple, 2013);
- multisector real business cycle (RBC) models, which deal only with short-term fluctuations in capital and not the long-term steady state (Atalay, 2017; Eaton et al., 2016a; vom Lehn and Winberry, 2022); and
- national accounting (Barro, 2021; Corrado et al., 2009; Hulten, 1979; Koh et al., 2020).

ACCOUNTING FRAMEWORK AND MEASUREMENT ISSUES: EXAMPLES OF AN ISLAND ECONOMY

Ding first presented two simple stylized models of an island economy. Economy (A), which consumes only fish, is closed to outside markets. Wages are normalized to a numeraire, and the production function of fish depends on labor and access to a fishing rod. Fishing rods are an intermediate input, and the production function of fishing rods depends only on labor. This means that there are two sectors in the economy: a fish-producing sector and a rod-producing sector. In the stylized example, the labor value added is split evenly between the two sectors. The fish-producing sector will employ labor and rods (an intermediate input) at equal proportions. Fishing rods are intermediate goods as they decay perfectly in 1 year. The final output, gross domestic product (GDP), will be equal to the size of the labor force. The simple model confirms the prediction from Hulten (1978) that analyzes the productivity shock in rods on GDP. One can take the Dolmar weight, the output of fishing rods with respect to labor ($\frac{1}{2}L$ in this example), and divide by the total GDP (L). In this example, the result is an elasticity, between GDP and a productivity shock in fishing rods, equal to one-half (see Figure 12-1).

In Ding's second stylized model, Economy (B), fishing rods that last for 2 years are now considered capital inputs, but require twice as many workers. The distribution of labor will be the same as in Economy (A), meaning that the number of rods used will also be the same. The major difference is that only half of the fishing rods need to be replaced each period. This is represented in the model as a $\frac{1}{2}L$ investment, and $\frac{1}{2}L$ capital value added, rather than a $\frac{1}{2}L$ intermediate and no capital value added as in the first economy. The resulting GDP in the economy is then 1.5 times the size of the labor force, rather than L in the first economy (see Figure 12-2). Barro (2021) explores issues of double counting in GDP, and reports that GDP in real consumption would be double counted, but GDP in terms of

	Rods	Fish	Consumption	Investment	Output
Rods	0	$\frac{1}{2}L$	0	0	$\frac{1}{2}L$
Fish	0	0	L	0	L
Labor	$\frac{1}{2}L$	$\frac{1}{2}L$			
Capital	0	0			
Output	$\frac{1}{2}L$	L			

- GDP = L
- d log GDP = $\frac{1}{2}$ d log Productivity shock in Rods

FIGURE 12-1 Economy (A)—Fishing rods as intermediate goods.
NOTE: GDP = gross domestic product.
SOURCE: Presentation by Xiang Ding.

capital value added would not. This has implications for the implied Hulten elasticity, defined above. In terms of capital value added, the elasticity of a productivity shock with respect to GDP is one-third. This differs from the elasticity for real consumption (one-half) and it does not coincide with GDP when measured as the sum of labor and capital value added, which is two-thirds. This illustrates the issue an outside observer would encounter if they were trying to estimate these elasticities from the final input-output table.

Ding stated that this is not an artifact of the stylized model and that it has real-world implications. In 2008, the System of National Accounts made a shift from weighting most global R&D output as intermediate expenses, to weighting them as an equal share with gross fixed-capital formation. The resulting inferences and calibrations from these datasets are then potentially drastically altered. Further, capital is traded at the same frequency as intermediate goods.

THEORETICAL FRAMEWORK

Ding moved to describing the theoretical model that will be calibrated. Households have a standard intertemporal preference function with discount factor, β. The rate of return, or risk-free rate r_f, for capital comes from the Euler equation. There are j industries that use intermediates and capital as factors in a Cobb-Douglas production function with constant returns to scale. Capital evolves in a standard way; capital tomorrow depends on depreciation, with rate δ_j, capital today plus investment today.

	Rods	Fish	Consumption	Investment	Output
Rods	0	0	0	$\frac{1}{2}L$	$\frac{1}{2}L$
Fish	0	0	L	0	L
Labor	$\frac{1}{2}L$	$\frac{1}{2}L$			
Capital	0	$\frac{1}{2}L$			
Output	$\frac{1}{2}L$	L			

- "GDP" $= 1.5L$, (not L)
- d log "GDP" $= \frac{1}{3}$ d log Productivity shock in Rods, (not $\frac{1}{2}$, nor $\frac{2}{3}$)

FIGURE 12-2 Economy (B)—Fishing rods as capital goods.
NOTE: GDP = gross domestic product.
SOURCE: Presentation by Xiang Ding.

Ding presented the key insight offered by the model as the link between investment flows and capital value added. The value of the net present asset is the sum of the rental price for the asset and the depreciated value tomorrow, discounted to today's prices. The steady-state predictions are that the value of the net present asset is the same today as tomorrow, or a constant. Further, investment is set at the level that replaces the depreciated capital stock. There is a no-arbitrage condition that pins down the return of risk-free capital, K_j, and the price, P_j, of a new investment good. This leads to the final equation,

$$P_j I_j = \frac{\delta_j}{r^f + \delta_j} r_j K_j ,$$

which relates the level of investment, I_j, to the return payments to capital for each asset j. The conditions for the steady-state equilibrium are standard: labor market clearance, asset market clearance, investment of each asset type achieves the required rate of return, and output market clearance. The output market clearing condition provides a deeper intuition. The total output produced by industry j is the sum of demand for intermediate inputs, the sum of expensed goods and new capital formation, and the demand for the final good, the sum of the labor income, and capital gains income.

Ding discussed one proposition of the model, the propagation of a shock in autarky, or the considered elasticities from the island economy. Using the Leontief inverse, the productivity shock in any industry can be related to the per-period steady-state welfare measured as the real final consumption. The inverse depends on the expenditures on intermediate inputs and implicit expenditures on capital goods. Domar weights (gross output over GDP) are no longer a sufficient statistic. The weights will now be modified because of the capital gains income that is consumed in the same period. This implies that the weights will not be modified only in the limit that the rental rate of capital goes to zero.

Ding explained that the model predicts that welfare gains of liberalization will be doubled if a country moves from autarky to an open economy. In a one-sector economy with a typical intermediate input share of 0.5, the proposed elasticity for the propagation of a shock with respect to real consumption is 2. In a world of intermediates and capitalized investments, the share moves from 0.5 to 0.75, providing an elasticity of 4.

The model in an open economy offers the same intuition as the above closed economy, but with the inclusion of international products and capital.

QUANTITATIVE RESULTS

Ding briefly discussed the calibration of the model. He used the 2007 release of the WIOT merged with socioeconomic data for the share of labor in each industry, using 33 countries and 33 industries. For the industry-level trade elasticities, he used three sources: Caliendo-Parro and Costinot and Rodriguez-Clare for trade in goods, and a constant elasticity of 5 for trade in services. The

WIOT offers information on labor incomes, gross output, import shares, intermediate input use, cost expensing (from supply and use), and cost civilization (only total value added from capital). To fit the model parameters, the linking equation discussed in the theory section is used, in connection with the BEA's Capital Flows Tables (see Meade et al., 2003) to decompose the capital value added. Ding used BEA estimates of depreciation rates by asset and assumed a global risk-free return to capital as the risk-free U.S. interest rate.

Following model calibration, Ding ran a counterfactual in which trade is liberalized by 10 percent across all countries and industries. Using the standard hat algebra approach (see Dekle et al., 2008), he solved for the exact welfare changes, which are the sum of the real labor income and transfer payments and real capital gains. The welfare impacts are then compared in two models calibrated to the same data: a setting with capital as the primary input and investment as final use, as in past literature, and a model with rental services from capital assets as intermediate inputs, which is Ding's contribution (see Figure 12-3).

The model predicts a doubling, an average factor of 2.4 across all industries, of the welfare gains of trade due to a 10 percent liberalization in the second model approach relative to the first, in both open and closed economies. Welfare gains are heterogeneous across countries, as seen when decomposing the gains by capital is final, real labor income, and real capital gains.

Ding stated that countries with high capital intensity, when capital is a final good, experience larger increases to their welfare following liberalization.

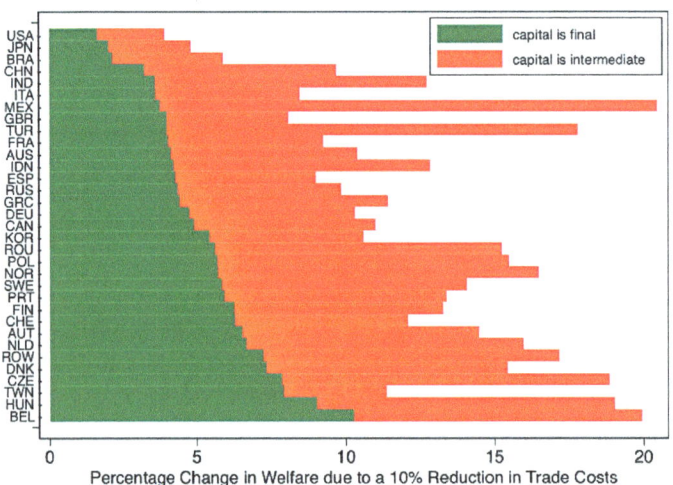

FIGURE 12-3 Gains from trade: Comparison across models.
SOURCE: Presentation of Xiang Ding.

The gains are also heterogeneous across industries in which capital-intensive industries benefit the most (see Figure 12-4).

SUMMARY

The measurement of supply and use of capital assets across industries using current macrodata is difficult, driving a disconnect between the reality of GVCs and the calibration of macro trade models. Ding offered a conceptual framework, in the form of a dynamic general equilibrium model, that treats rental of capital as an intermediate expense. This model offers comparable steady-state predictions to static models and follows a similar solution method, hat algebra, with a slightly altered calibration. The model predicts that treating capital as an intermediate good doubles the welfare gains from liberalization with the largest impact on capital-intensive countries and industries.

Ding mentioned several channels for future work in this area, adding to the theory for adjustment to new steady states, the implications for measurement in national accounts, the role of multinational firms and foreign direct investment, and profit-shifting incentives for tax avoidance.

DISCUSSION

Discussant: Brent Moulton (International Monetary Fund)

Brent Moulton, senior economist at the International Monetary Fund, began by giving a summary of the paper and contextualizing it in the existing

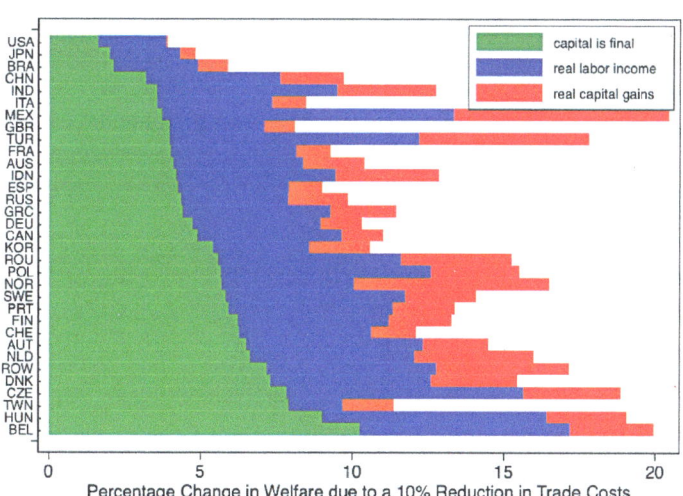

FIGURE 12-4 Gains from trade: Comparison across models.
SOURCE: Presentation of Xiang Ding.

literature. Prior work in this area focuses on the use of global suppliers to transform intermediate inputs to final goods. The paper expands this view of GVCs to include capital inputs by treating them analogously to traditional intermediate inputs. Relative to the prior models that do not treat capital as an intermediate input, the paper reports that their inclusion more than doubles the gains from trade in terms of real consumption. These gains are strongest in capital-intensive countries.

Moulton focused his comments on the measurement issues raised rather than the trade theory, offering several main comments before moving to a deeper discussion of the measurement issues that Ding raised. As the paper is in an early stage, there are several avenues through which it can increase its contribution to the field. The paper would benefit by connecting with the productivity literature, which provides solutions to several measurement concerns raised. This literature includes several datasets that could be incorporated into the paper. Moulton's last main comment is on the terminology used; one should be careful in the comparison of capital and traditional intermediates as they behave in separate manners. More work on developing the framing of this work would also be useful.

Moving to the measurement of capital services, Moulton offered some background on the national accounts data. The founders of national accounting (Richard Stone, Simon Kuznets, and others) treated capital as a consumable in the production process and noted the services that capital provides. Their terminology, *intermediate consumption for rented capital*, and *consumption of fixed capital for owned capital*, indicate this understanding. However, practical measures for differentiating capital goods and their respective services were not available in the 1930s and 1940s, so capital goods were treated as gross operating surplus.

Moulton went on to say that economists in the 1960s and 1970s studying growth accounting and productivity produced measures of capital services that differentiate various forms. Some of the more notable papers in this literature are Jorgenson (1963) on capital theory and investment behavior; Jorgenson and Griliches (1967), which explained productivity change; Hulten (1973), which introduced Divisia index numbers; Diewert (1976), which introduced exact and superlative index numbers; and Hulten (1978) on growth accounting and intermediate inputs.

Moulton explained that the methods developed in the prior decades were implemented in the 1980s in studies on multifactor, or total factor, productivity of the macroeconomy and its constituent industries. In 1983, the Bureau of Labor Statistics began their analysis of trends in multifactor productivity with work documenting the trends between 1948 and 1981. Gollop and colleagues (1987) offered a major study of productivity in the United States. These authors aggregated capital services for multiple capital-asset classes with intermediate inputs and labor services. In 2001, the Organisation for Economic Co-operation and Development published a handbook on the measurement of productivity at the macro and industry levels. Moulton suggested that this work would be particularly interesting and beneficial in the context of the paper for the understanding of industries. Especially of interest are such industries as information and computing technology services that experience large price changes. Lastly, the System of National Accounts (SNA) (European Commission et al., 2009) has a chapter on the role of capital services.

Moulton transitioned to discuss the measurement of capital flows. The paper has two data sources: the WIOTs, which made a change in 2008 regarding the relative importance of capital and traditional intermediates, and the U.S. BEA capital flow tables for 1997. The BEA publishes alternate estimates of investment in detailed capital assets by industry, which are not as detailed as the 1997 tables but still have large coverage and may be useful. The more recent BEA estimates were improved by a more detailed collection of capital asset classes during the Census Bureau's annual capital expenditure survey.

Moulton pointed out that international capital flow tables, such as those developed by the BEA, are not in the 2008 SNA, and many countries have not developed them. Several countries have developed these estimates as part of their multifactor productivity accounts, or estimates were developed by university research teams. The KLEMS (analysis of capital [K], labor [L], energy [E], materials [M], and service [S] inputs) project, organized by Jorgenson, Timmer, and Van Ark, includes capital-flow estimates for the European Union, United States, China, India, Canada, and several other countries.

Moulton's final comment was directed at Ding's treatment of capital assets as an intermediate input. There are analogies between the capital goods and traditional intermediate goods, but there are also several differences. The SNA does not treat capital assets as intermediate inputs, as they are owned by the firm and semidurable. Further, capital assets have value and are recorded on firm balance sheets. No direct measure of rental service or income flow from production exists; so, for practicality, it is simpler to record income from capital assets as a segment of the operating surplus.

The productivity literature has developed methods for and theory for measuring capital services, and Moulton recommended their introduction into the paper. The paper relies heavily on steady-state conditions to connect capital flows and services. For this reason, Moulton suggested that Ding think more about how capital flows vary in the temporal dimension when the conditions for a steady state are not met. For example, a company has a 5-year-old asset that, if replaced today, would be sourced from Thailand; but since the asset has 5 more years of life, it may be sourced from the Philippines when they choose to invest in new equipment. In this case, how would this flow be measured: the observed flow as purchase, an imputed flow at time of use, or the expected flow at replacement?

One audience member asked how to reconcile and explain the gains from trade results from the input-output tables in the more complex quantitative tables and the stylized tables in the example. Ding responded that he keeps the use of intermediate input and simply shifts capital value added into a vector across all asset classes. He retains the standard input-output linkages, but adds in new ones based on the capital intensity of an industry.

13

Colocation of Production and Innovation: Evidence from the United States

Paper Authors: Teresa C. Fort (Dartmouth College), Wolfgang Keller (University of Colorado Boulder, National Bureau of Economic Research [NBER], and Centre for Economic Policy and Research [CEPR]), Peter K. Schott (Yale School of Management, CEPR, and NBER), Stephen Yeaple (The Pennsylvania State University and NBER), and Nikolas Zolas (U.S. Census Bureau)

Presenter: Teresa C. Fort (Dartmouth College)
Moderator: Justin Pierce (Federal Reserve Board)

The workshop paper by Teresa Fort, Wolfgang Keller, Peter Schott, Stephen Yeaple, and Nikolas Zolas studies how firms colocate production and innovation in the context of the United States.

Teresa C. Fort, associate professor of business administration at the Tuck School of Business at Dartmouth College, introduced the workshop paper by highlighting policy questions about the potential impact of the decline in manufacturing employment in the United States on U.S. innovation. There is concern that the manufacturing loss in the United States will reduce U.S. innovation, either because of increased competition from imports and the resulting decline in domestic research and development (R&D), or because of complementarities in the colocation of innovation and production. Bloom and colleagues (2020) show that innovative efficiency is falling and that maintaining similar levels of innovation now requires more workers than in the past.

Fort described the three main questions the paper is intended to address:

- How have the characteristics of innovating firms evolved with time?
- Is R&D effective only if it is colocated with production?
- What are the mechanisms behind the relationship between production and innovation?

Manufacturing declines in the United States may affect domestic innovation in one of two ways:

- Complementarities from the colocation of production and R&D may exist that allow innovation to occur more efficiently when located near the production site, meaning declines in domestic manufacturing decrease innovative activity in the United States.
- Alternatively, innovative activity may benefit from reallocation and specialization due to comparative advantage effects, which would benefit the United States. The problem of colocation has two dimensions: geographic and firm borders, and their interaction.

CONTRIBUTION

Fort explained that the paper's first contribution is a set of stylized facts on the trends of innovating firms in the United States from 1977 to 2016. Patenting activity moves from manufacturing firms (MFs) to nonmanufacturing firms (NMFs). An MF is defined as a firm that has at least one manufacturing establishment in a given year in the United States. MFs account for a vast majority of patenting activity early in the sample, decreasing to about half at the end of the sample. The authors focused on a subset of NMFs, those that were once manufacturers but are no longer, which they called former manufacturing firms (FMFs). FMFs that transitioned from MF to NMF later in the sample continue to innovate, which suggests that colocation is not as important for this subset.

As Fort described, firms that contain both manufacturing (M) and innovation (P) establishments patent more relative to other firms. The distance between M and P establishments within a firm tends to increase over time, but some remain close. The firms that locate M and P establishments within 5 miles of each other are found to patent about 12 percent more.

Fort explained that she and her coauthors plan to analyze where innovative activity occurs within firms and understand the margins that alter the colocation decisions of firms. In addition, they plan to causally identify and understand the relationship of innovation and distance across firms.

DATA

Fort described the U.S. Census Bureau datasets used for this study:

- The Longitudinal Business Database (LBD), which contains all private, nonfarm establishments from 1977 to 2016. Establishment-level North American Industry Classification System [NAICS] classifications and establishment geocodes from the Business Register (from Fort and Klimet, 2018) were added for all establishments in the dataset.
- The economic censuses of manufacturing wholesale, retail, and services firms provided establishment-level sales and inputs data at 5-year increments.

- The Longitudinal Foreign Trade Transactions Database provided firm-level import and export transactions from 1992 to 2016.
- Two R&D surveys provided information on innovative activity of firms from 1977 to 2016.
  - Survey on Industrial Research and Development (1977–2007)
  - Business R&D and Innovation Survey (2008–2016)
- The U.S. Patent and Trademark Office (USPTO) database provided the names and addresses of firms matched to the LBD and identified manufacturing and processing patents from 1973 to 2018, an extension of the prior matched database.

PORTRAIT OF U.S. INNOVATION

Fort showed that innovation activity in the United States, measured by patents granted and R&D expenditures, has grown in the 40-year span of 1977 to 2017 (see Figure 13-1). She and her coauthors found a decline in activity after 2015, which is an artifact of the application-to-grant lag.

Fort then explained that she and her coauthors categorized firms into three mutually exclusive types for each year: MFs, at least one manufacturing plant in year t; NMFs, no manufacturing plants in year t; and FMFs, a firm that was an MF prior to year t. NMFs dominated the distribution of firms both in the number of firms and number of employees. The number of MFs was also flat across the sample. Employment in NMFs grew from 1977 to 2017, while employment in MFs, including both manufacturing and nonmanufacturing employees, was relatively flat. The share of employment in MFs classified as manufacturing decreased over the sample. FMFs grew in number and employment over the sample.

The share of innovation by MFs declined from 91 percent in 1977 to 54 percent in 2016, but the MFs were still innovating at nearly double the rate in

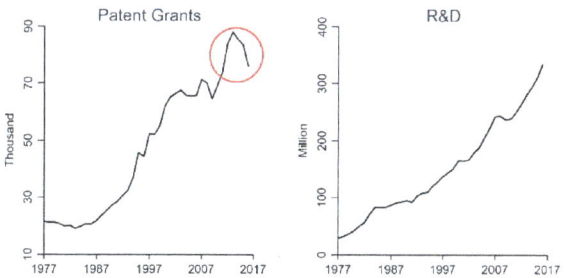

FIGURE 13-1 U.S. innovation growth over the last 40 years.
NOTES: Granted patents are assigned to application year. R&D = research and development.
SOURCE: Presentation by Teresa Fort.

2016 as they did in 1977 (see Figure 13-2). In 2016, NMFs accounted for 28 percent of patents and FMFs accounted for 18 percent of patents. This implies that a sizable number of firms are capable of innovating without operating a manufacturing establishment. R&D expenditures show similar trends.

Fort said that permanent FMFs, or firms that exit manufacturing permanently, were split into five cohorts: 1978–1996, 1997–2001, 2002–2006, 2007–2011, and 2012–2016 (see Figure 13-3). Firms in the 2007–2011 cohort showed the strongest growth in terms of innovative activity, whereas the 2012–2016 cohort displayed the second-highest innovative activity, although this growth was muted relative to the 2007–2011 cohort. The 2007–2011 cohort also dramatically increased imports from China over the sample, in absolute terms and relative to the other cohorts. This suggests that these firms, predominantly in computer and electronics manufacturing, likely moved production offshore while continuing to innovate in the United States.

The main takeaways, according to Fort, are that NMFs accounted for an increasing share of patents and R&D expenditures over the sample; FMFs continue to patent, especially firms that recently exited manufacturing; and finally, FMFs that exited manufacturing more recently show a large increase in imports from China, indicating offshoring of production.

MEASUREMENT OF COLOCATION

Fort explained that she and her coauthors developed new measures for colocation of U.S. production and innovation. They identified establishments within a firm that focus on innovation (P establishments) from the following

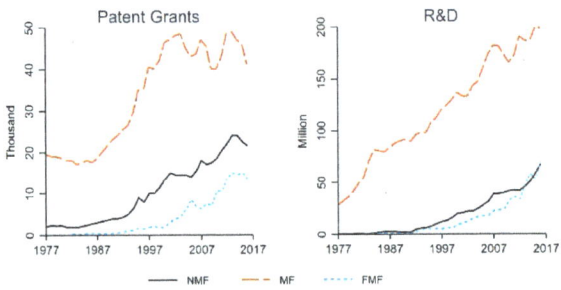

FIGURE 13-2 Manufacturing firms dominate U.S. innovation.
NOTES: MF patent shares declined from 91 percent to 54 percent in 2016. NMFs accounted for 28 percent of patents in 2016. FMFs accounted for 18 percent of patents in 2016. FMF = former manufacturing firm; MF = manufacturing firm; NMF = nonmanufacturing firm; R&D = research and development.
SOURCE: Presentation by Teresa Fort.

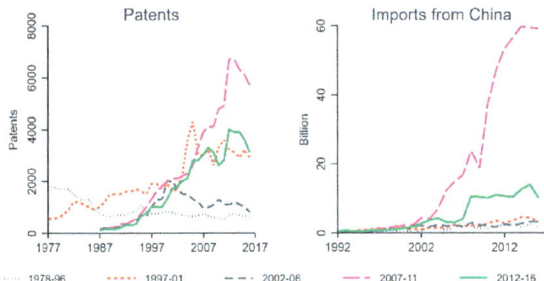

FIGURE 13-3 Patents by permanent FMFs differ by cohort.
NOTES: Firms that exited manufacturing from 2007 to 2011 exhibited the strongest growth. The 2007–2011 cohort also exhibited dramatic growth in Chinese imports. Employment dynamics were similar across cohorts.
SOURCE: Presentation by Teresa Fort.

NAICS codes: establishments that focus on scientific R&D services (NAICS 5417); professional scientific and technical services (NAICS 5413–5416); corporate, subsidiary, and regional managing offices (NAICS 551114); and information and telecommunications (NAICS 5112, 517, 518).

Fort and her coauthors found that firms with both M and P establishments patent 65 percentage points more than firms that lack M or P establishments. When including firm fixed effects—therefore looking at within-firm effects—firms with both M and P establishments patent 15 percentage points more than firms that lack M or P establishments. The authors removed variation from time-varying firm size, age, and stocks of patents.

To measure the distance between manufacturing and innovating establishments, Fort and her coauthors focused on firms with both M and P establishments (MP firms). The average and minimum distances between M and P plants was calculated. According to Fort, the median of the average distance between M and P firms was hundreds of miles across the sample, and this distance grew over time from 301 miles in 1977 to 416 miles in 2012. This may suggest that there is no colocation in the United States; however, the median firm has at least one pair of M and P establishments that are colocated within 6 miles of each other.

RESULTS

Fort and her coauthors estimated a relationship in which the dependent variable is the inverse hyperbolic sine transformation of the sum of subsequently granted patents (or citations) applied for by firms in year t to $t+4$. This specification allows the results to have an elasticity interpretation. The explanatory variables are the minimum distance, split into two subsets: 0–5 miles and 5–60 miles. The depreciated and 1-year lagged patent stock and a set of controls that account for time-varying differences in firm size and age, along with year, county, and firm fixed effects are used to remove unwanted variation in the empirical results.

Across all outcome variables of interest (patents, citations, manufacturing patents, and process patents), firms with at least one set of M and P plants within 5 miles of each other innovated more. This relationship also held for firms with a set of M and P plants between 5 and 60 miles of each other. However, the coefficient on patents, which is the measure of innovation, declined from 11.6 percentage points to 7.6 percentage points when the colocation distance increased. According to Fort, this suggests that colocation does matter. At the same time, she explained that the mechanisms affecting innovation are not clearly understood when there is colocation of production and innovation, and there may be something special about the management of firms that have both manufacturing and innovation establishments.

FUTURE WORK

Fort then discussed future extensions of her and her coauthors' work. They would like to see whether patent creation occurs in a firm's colocated region; in other words, does innovation at the P plant influence behavior at the nearby M plant? Or is innovation influenced by behavior at another M plant that is not nearby? They will be looking at patenting behavior in a region to understand the relationship between innovation and manufacturing, and to identify any spillover effects from other firms' M and P plants in a region.

According to Fort, she and her coauthors would also like to understand policy-induced changes to the cost of R&D, such as how a measure of R&D tax credits interacts with manufacturing and innovation establishments.

SUMMARY

Fort concluded by summarizing the new stylized facts from her and her coauthors' work:

- The share of aggregate patents by NMFs increased from 9 percent in 1977 to 46 percent in 2016. This suggests that manufacturing is not essential for continued innovation, even in manufacturing-related activities.
- Firms with both manufacturing and innovation establishments innovate more. The spatial distribution of firms is broad across the establishments, but firms' patents are higher when the establishments are colocated. In future work, they will explore the mechanisms that drive this pattern, particularly around the influence of the innovation establishment on the behavior of local manufacturing and the productivity of innovation in colocated regions.

DISCUSSION

Discussant: Nick Bloom (Stanford University)

Nick Bloom, William D. Eberle professor of economics at Stanford University, began by placing the paper into two important trends in the U.S. economy: productivity growth is slowing and manufacturing share of employment is falling. Productivity growth is down across all countries in the Organisation for Economic Co-operation and Development—essentially across all developed countries. While manufacturing output is flat, the manufacturing employment share of total employment has been consistently decreasing since 1945. Historically, innovation was concentrated in the manufacturing sectors, so one may think that these trends are connected; the authors sought to understand if this is the case. These questions are relevant not only to the academic audience, but also to policy makers, financial markets, and the general public.

Bloom offered a summary of the work, starting with a brief discussion of the data. The authors combined several U.S. Census datasets with the novel contribution of patent and geocode data. Bloom emphasized the amount of work that went into this process and stated that this work is relevant to policy questions currently being asked about innovation, such as those related to global value chains for solar panels, with respect to China and the United States. Additionally, he noted that the Biden administration is pushing to shift manufacturing back to the United States because it believes that it is important for innovation.

The paper offers four key findings:

- First, the stylized fact that NMFs account for 46 percent of total U.S. patents in 2016, an increase from 9 percent in 1977. The FMFs are a part of this shift of innovation away from manufacturing, which is interesting as these firms continue to locate their headquarters in the United States, while shifting production to foreign countries. This is consistent with the "coastal" story of such firms as Apple, Levi, Hasbro, or Dell—innovating in the United States and producing abroad while achieving high profit rates. The most innovative U.S. firms are primarily technology firms, which suggests that manufacturing is not the only path to innovation.
- Second, FMFs that shifted production to other countries continue to innovate in the United States.
- Third, the firms that have both manufacturing and innovation establishments nearby, or colocated, appear to patent much more than firms that do not. Firms that have only manufacturing plants patent 4 percentage points more, firms with only innovative plants patent 2 percentage points more, and firms with both patent 66 percentage points more, suggesting strong colocation effects, the opposite of the coastal story.

- Finally, firms with manufacturing and innovation plants in close proximity are more likely to patent. If the plants are within 5 miles of one another, then patenting increases by about 11 percentage points. For firms between 5 and 60 miles of each other, the increase is about 7 percentage points, though this measurement is sensitive to the specification.

Bloom suggested that the key question remaining is about the underlying mechanisms at play. Do these results suggest colocation effects (colocation causing more innovation), or are they a consequence of other firm characteristics, such as high-quality management? Identifying the correct mechanism has policy implications. If there are serious colocation effects, then domestic policy makers may have an incentive to introduce tariffs and subsidize manufacturing. If innovation is independent of manufacturing, then policy makers have an incentive to invest in domestic R&D and education. The fact that the Biden administration is currently discussing these policy questions indicates the importance of this paper in nonacademic circles.

Fort's response focused on two points raised in the discussion:

- The results regarding FMFs and colocation effects appear to contradict each other. The FMFs continue to patent, but firms with both manufacturing and innovation plants patent much more. One fact that may help reconcile this is that for firms that colocate within 5 miles, manufacturing employment for these collocating firms has fallen precipitously over the sample period.
- Second, Fort disagreed with the Bloom's policy suggestion to use tariffs and subsidies to incentivize domestic manufacturing, even if colocation effects are determined to be strongest. Firms that are shedding manufacturing are not targeting colocated manufacturing plants, but rather their noncolocated plants—likely the larger, labor-intensive plants. The plants that benefit from colocation remain. She sees across-firm spillovers as the more important policy dimension.

One audience member asked if the FMFs are altering the type of innovation in which they participate. Fort responded that the authors have looked at this and the firms appear to continue to patent in the same innovation type. Another audience member wondered if firms outsource manufacturing domestically. Fort responded that this intuition is correct. In the 2007 data from the Bureau of Economic Analysis (BEA), there are no firms that produce outside of the United States that do not also produce within the United States. Another audience member wondered if the authors have looked at the variation of geographic region in the data. Are the manufacturing and innovation plants both on the coast or are they in less attractive locations? Fort responded that this is an interesting idea and something that the authors will investigate.

Another audience member wondered if overseas R&D centers of U.S. multinational enterprises appear in the data and if nonpatenting innovation appears. Fort responded that in their findings, a shift in innovation towards nonmanufacturing firms would be even stronger if they were able to include nonpatenting innovation. Nikolas Zolas, one of the paper's coauthors, has worked on alternate measures of innovation, such as trademarks, that are more common among the service sectors. In future work, the coauthors may try to include this measure in their analysis, as well as other measures from the BEA censuses. Another audience member wondered what role mergers and acquisitions play in the colocation results. Fort responded that this is a good idea and something that the authors will investigate.

14

Global Value Chain Measurement Methodology: Challenges and Prospects

Sally Thompson (former deputy director, Bureau of Economic Analysis [BEA], retired) moderated a session on methodological issues in measuring global value chains (GVCs). The session included three speakers: Timothy Sturgeon (senior researcher, Massachusetts Institute of Technology Industrial Performance Center) on GVC measurement issues; Oscar Lemmers (senior researcher, Statistics Netherlands) on challenges in modernizing official business statistics; and Jon Samuels (senior research economist, BEA) on measuring research and development (R&D) and competitiveness for U.S. value-added exports. Francisco Moris (senior analyst, National Center for Science and Engineering Statistics [NCSES], National Science Foundation [NSF]) served as discussant.

Before introducing the speakers, Thompson noted her sadness at the passing of Ray Mataloni, head of the research program in international economics at BEA. He was an expert in multinational enterprise (MNE) data and their use in assessing the economic impacts of globalization and would have very much enjoyed the workshop.

MEASURING GLOBAL VALUE CHAINS

Timothy Sturgeon noted that he was not presenting a paper per se, but rather remarking on key issues he believes are important for the field and referencing relevant papers prepared for the workshop. He has worked on the idea of GVCs for about 25 years and on the measurement of GVCs since about 2006.

Effects of Global Value Chains on the Global Environment

Sturgeon said that our model of trade has historically been based on competitive advantage and foreign direct investment (FDI). Less-developed countries allowed FDI but often had local content requirements. Starting in the late 1980s and accelerating in the 1990s, especially with China's accession to the World Trade Organization in 2001, a changed global environment took shape with some special features. These features included vertical specialization in trade and FDI and, starting in the mid-2000s, a push into international trade in services in GVCs; outsourcing and offshoring continued, particularly to India. Most recent changes have included the growth of knowledge and innovation networks and the global fragmentation of R&D (see Gumpert et al., Chapter 2).

The changing global trade environment has, in Sturgeon's view, complicated matters for policy makers. There are fewer clean industrial policy tools for either developing or advanced countries, whether on the lead firm side or the supplier side. Consequently, policymakers have a big challenge to understand industry value chains in detail. Moreover, industries behave differently in the GVC context, so the demands on the data side are high.

The world has moved from fixed comparative advantage to dynamic comparative advantage, Sturgeon said. Data resources are strong for analyzing trade in goods but weak for trade in services, enterprise characteristics, and intangible assets, although there are some innovative ways to tackle measurement.

Global Supply Chains

Sturgeon defined supply chains as distinct from GVCs. Any "lead" firm controls a supply base—for example, Ford with its supply base competes against Toyota with its supply base in the United States or internationally. In contrast, the value chain is the whole chain, up to the end users and even farther to recycling—what is known as the circular economy. See Figure 14-1 for a schematic.

While the global economy is a network or a series of nested networks, Sturgeon emphasized the importance of analyzing lead firms, which he defined as not necessarily the largest firms or the best firms but firms that initiate the activities of chains. In other words, lead firms have their own supply chains, as do specific products.

Sturgeon explained that, because lead firms have supply chains, they decide who is in and who is out, giving them buyer power, which is an important role. They also add value of their own in R&D and innovation, and they commonly capture the lion's share of the rent. Oddly, though, the supply-management literature is focused on the strategy of the lead firm for supply chains and other phenomena but not the added value of the lead firm. On the supplier side, the literature looks at what suppliers are doing and what value they are adding, but not their strategy. Suppliers are trying to make a profit in this environment, and while the literature is very weak on supply chain management, more developed literature might help suppliers know what to do.

Sturgeon noted a key finding in the literature—in the 2000s, lead firms' outsourcing led to shared supply bases and the emergence of global suppliers. Prior to this period, big firms occupied the front of the value chain and in capital-intensive pieces, such as materials (e.g., steel), but smaller firms made up other

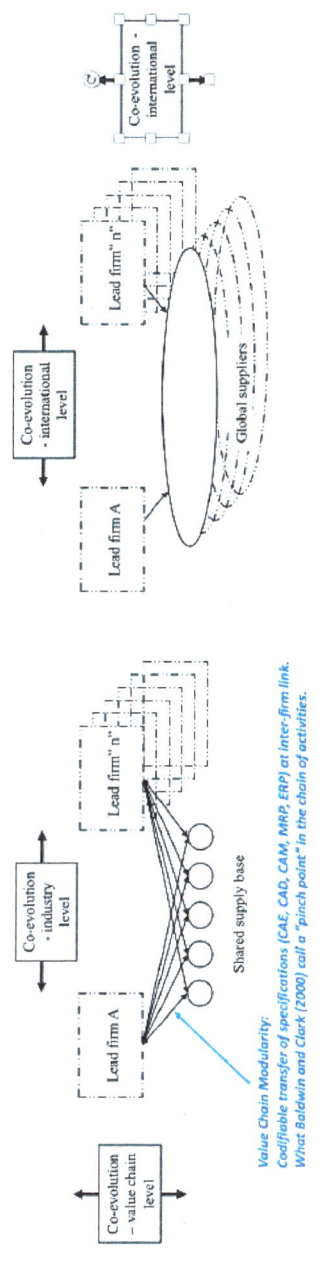

FIGURE 14-1 Global value chains: Stylized narrative—From outsourcing to outsourcing *and* offshoring.
NOTES: FDI = foreign direct investment; ICT = information communication technology; MNC = multinational corporation.
SOURCE: Presentation by Timothy Sturgeon.

links in the chain. That is no longer the case. Working with regional supply bases proved too complex for lead firms, as they went offshore in search of markets and low-cost export platforms. Consequently, there are now huge multinational firms at each stage of the value chain.

Global Value Chains

Sturgeon said there are different perspectives on what constitutes a GVC. The value-added perspective looks at long production chains that cross more than one border. The researchers with whom Sturgeon works ask what happens in a given transaction: Is explicit coordination needed? Is there customization? Are there relationship linkages or modular linkages? Or are the transactions arm's length?

To address these questions, Sturgeon continued, the field tends to focus on specific intermediate trade flows in complex assembly industries with long supply chains in which explicit coordination is required of the parties. These industries include apparel, electronics, motor vehicles, and commercial aircraft, each having a GVC with a very different shape. Primary commodities—that is, products sold at auction or with an international reference price—are not involved in GVCs.

Sturgeon identified as an important phenomenon the geographic separation of innovation from production, which has led to different outcomes over time in high-income economies (e.g., the United States) compared with middle-income, industrializing, GVC-linked economies (e.g., China). The former exhibit a high share of value capture and control over the industry architecture and trajectory, with de facto standard-setting and buying power; they also exhibit high-market-cap industries with relatively small employment and continued or accelerated secular decline in manufacturing employment. In contrast, such economies as China's exhibit a low share of value added in GVCs, exclusion from innovation functions and standard setting, employment creation in manufacturing, and a two-tier manufacturing structure for export versus domestic consumption. Such economies can use their foreign-exchange earnings from exports to invest in domestic innovation capacity, as China is trying to do. However, Sturgeon said the jury is out on whether that can actually work. The Chinese semiconductor industry has not gained a substantial portion of market share, despite investing billions of dollars and acquiring equipment and expertise from abroad.

Sturgeon pointed out that, while research tends to focus on bilateral and firm-level dynamics, such as reallocating resources from manufacturing to innovation, this focus can miss industry-level dynamics in which there are positive external-scale economies that build up at the industry level. This focus can also undervalue the innovation and standard-setting benefits from geographic specialization and from supportive domestic institutions and government. There is a reason that Silicon Valley continues to lead in wave after wave of innovation, because those kinds of institutions and standard-setting power are located in the Valley.

Sturgeon noted another phenomenon of interest—geographically embedded capabilities evolve over time, and the geographical specialization evident in GVCs appears to be path dependent. One of Sturgeon's current projects is looking at the mobile phone handset industry, in which the component supply chain is becoming concentrated in fewer companies, so that it is very geographically specialized. In each geographic specialization, there is a concentration within the vertical segment, creating vulnerabilities in these supply chains as they are dispersed but concentrated in each node such that they are indispensable. For something like a mobile phone handset, substitution is difficult if not impossible; that kind of vulnerability in the supply chain is easy to underestimate and can be exploited or even weaponized. The COVID-19 pandemic revealed the fragility of many supply chains, which were so finely tuned, dispersed, highly efficient, and low cost that there were no buffers or alternative sources. The policy focus on GVCs and supply chains is now intense.

The measurement challenges are obvious. Information on FDI trade by industry, stage in the value chain (primary, intermediate, and final), firm ownership, and knowledge-intensive services is essential but not available. Even lead firms do not know everything that is in their supply chain. So far, solutions to the data problems have been piecemeal. Eurostat's new GVC surveys (previously called international sourcing surveys), which examine distinct business functions, are the only recent addition to its program to modernize relevant data systems (e.g., Smeets and Warzynski, Chapter 10). Everything else has involved complementary groupings, microdata linking, or mining administrative data in a way to avoid adding cost or burden for respondents (see, e.g., Wang et al., Chapter 3; and Fort et al., Chapter 13). The leveraging of private and semiprivate data can also be useful, but these data are not necessarily a public resource, and research using them is not reproducible.

Sturgeon concluded by outlining a process by which the policy, statistical, and research communities can help drive the creation of more relevant and useful statistics (see Figure 14-2). He praised collaborations of researchers with the data-producing community, but asked whether data development and research has to be done on a 20-year cycle.

QUALITY CHALLENGES IN MODERNIZING
OFFICIAL BUSINESS STATISTICS

Oscar Lemmers said his organization, Statistics Netherlands, provides scientists with the data they need. He admitted that the cycle of determining needs and providing relevant data can be slow.

Lemmers said he would discuss globalization, reasons why R&D and GVCs are difficult to measure, and potential solutions for the measurement problems—always through the lens of data quality. In quality terms, the questions are whether one is measuring well what one wants to measure or something else, and whether one is measuring well something that is actually useful to the outside world. Of course, the answers to these questions all depend on the specific topics

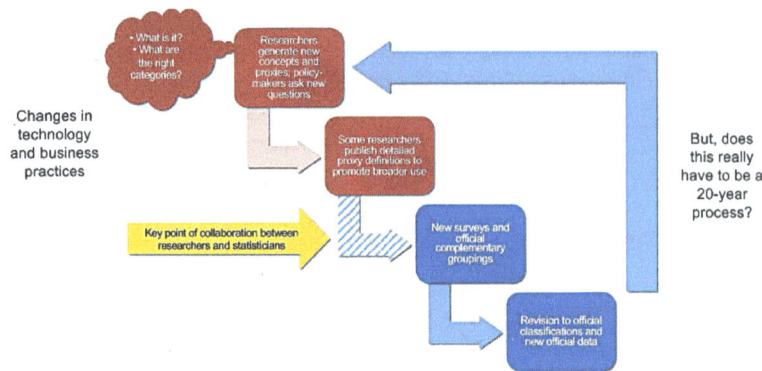

FIGURE 14-2 Policy, statistical, and research communities can help drive the creation of more relevant and useful statistics.
SOURCE: Presentation by Timothy Sturgeon.

at hand. In general, measurement is easier for one's own country. Researchers know their own country and the firms there. Firms may even be legally obliged to answer researchers' questions.

Lemmers stated that R&D as a share of gross domestic product (GDP) is measurable. Much more difficult is measuring how much R&D an industry (high or low tech) is using and how much R&D is coming from one's own country. He said one always needs to keep in mind that firm-level business statistics only allow us to measure the first step in the chain.

Lemmers revisited the history of R&D measurement. In the past, measurement was comparatively easy—production and R&D were in the same unit, and a statistical agency could send them a survey. Now measurement is more difficult because large firms can and do conduct production and R&D in different units. One way to deal with this problem is to talk with large and important firms to find out their structure and where to obtain information about their R&D. Then one can link microdata on R&D and production in a postprocessing operation.

Of course, many small firms do R&D as well, Lemmers noted, and the statistical agency must be concerned about coverage of the population. A solution could be the use of registers, when firms file for R&D tax credits, for example, or perhaps web scraping.

Lemmers observed that GVCs exacerbate measurement problems because production processes are often split up among units not only in the home country, but also among countries. Lemmers began with the simple case in which a domestic firm buys R&D from a different company in another country. This situation is measurable—there are paid bills for R&D inputs from a different firm. It becomes more difficult if R&D is offshored within the same company. It might appear that there is no R&D, when in fact it is happening abroad. Moreover, R&D from a parent or sister company can be obtained in different ways—not just in a measurable transfer on the account books, but also through licensing or royalties or something else. Lemmers also noted the role of tax incentives—where do companies want to book their profits? A statistical agency that wants accurate data on any of these aspects of R&D has to reengineer business data to fit the agency's statistical concepts in order to find out how much has been paid. On the other hand, it may prove infeasible to obtain appropriate data to fit the concept, which means the concept may be beautiful but not useful.

Payment for R&D that involves staff hours can be even harder to measure, according to Lemmers. He gave the example of writing a paper with someone abroad (he recently finished a paper with his Polish colleagues), when one may have no idea how many hours the other people put in. The difficulty is that the data get softer and softer when dealing with services. Lemmers believed there are ways to collect data for most questions, but not everything is feasible.

One approach would be to profile an entire firm worldwide using big data and other sources to understand the structure of the firm. Then the agency could tailor its survey—for example, the survey could ask about the type of R&D arrangements in addition to the value. The agency could also narrow down the population of firms and work with a smaller sample to open the black box and determine where the relevant information is located.

Lemmers believed that while this approach might still involve postprocessing of the data, it could work, although it might not be suitable for regularly issued official statistics because of its time-consuming nature. Researchers, however, could obtain useful information by working at the micro level. They could use an input-output model or look at detailed multiregion supply-use tables. With either approach, researchers would be examining more steps in the global chain, in contrast to business statistics, which traditionally cover only one step.

MEASURING R&D AND COMPETITIVENESS OF U.S. VALUE-ADDED EXPORTS: PROSPECTS AND CHALLENGES

Jon Samuels (BEA) said he would address measuring R&D and competitiveness of U.S. value-added exports. He wanted to emphasize the prospects for measurement, although he understood there are significant challenges. His basic thesis—or major prospect—is that linking value-added trade and KLEMS data[1] has the potential to yield important insights into the origins of GVCs and international competitiveness. Samuels asserted that familiar value-chain methods would allow parsing U.S. value added in GVCs, and KLEMS methods would allow identifying the R&D content that is embedded in U.S. value-added exports. Samuels also thought this approach would help identify the origins of economic competitiveness along GVCs. He noted that his work is in a preliminary stage.

---

[1] KLEMS (K-capital, L-labor, E-energy, M-materials, and S-purchased services) refers to intermediate inputs to the production of goods and services.

Stylized Example: Estimating R&D in $1.00 of Farm Exports

Samuels presented a stylized example of his approach for measuring the R&D content of U.S. exports, using an export of farm product. At first glance, it appears that there is no R&D content in this export, but looking closer at the embedded R&D in the intermediate products needed to produce farm product, such as fertilizers, gives a more accurate measure of the amount of R&D that is embedded in the export. This example helps illustrate how to calculate the full value of R&D in exports by including the embedded R&D in the inputs that go into the exported good.

- Every $1.00 of farm product exported from the United States requires production of $1.25 of product by U.S. farmers and the production of $0.95 of other domestic goods and services that feed into this farm production; the value added by the farm industry in this $1.25 of corn produced, and ultimately in the $1.00 of corn exported, is $0.43 (based on BEA's domestic requirements table).[2]
- However, the KLEMS shares indicate that there is $0.00 R&D in the $0.43 value added (from the BEA-BLS Integrated Industry-level Production Account, which gives the capital service share of R&D), giving the impression that there is no R&D content in the export of farm product.
- The solution is to go back to the $0.95 of other goods and services that goes into exporting $1.00 of corn and to decompose this into the value added contributed at each stage of production.
- One of the major inputs is chemical production—$0.07 of chemical-sector input is required, of this $0.03 is value added from the chemical sector (from the domestic requirements table), and $0.014 of this $0.03 is a payment to R&D capital services (from the BEA-BLS Integrated Industry-level Production Account).
- Accounting similarly for all the other intermediate inputs from industries that feed into the agricultural sector gives an estimate of $0.024 embedded R&D from U.S. producers in the $1.00 of exported corn.

Datasets that Enable Estimation of Value-Added R&D

Samuels then turned to the datasets used in the decomposition in his stylized example. The first dataset is the domestic requirements tables, which BEA has published since 2017. They split out imports from total requirements and give the amount of domestic intermediate inputs required, both directly and indirectly, for industries to supply goods and services to final demand.

---

[2] Samuels clarified that $1 of corn for export requires more than $1 of corn to be produced because corn is also used as an intermediate input in corn production.

While most R&D is not an intermediate input, Samuels said, it is important to know its role in production and how much is embedded in value added and gross exports. This is where the KLEMS statistics come in. This dataset is jointly produced by BEA and the Bureau of Labor Statistics. It includes estimates of R&D's role in production at the industry level. To measure the R&D component of capital services, one starts with information on how establishments invest in R&D capital stock. The R&D capital stock in turn yields capital service flow into production. The value of these capital services is basically quantity (the productive stock) times price: the perpetual inventory method applied to past R&D investments times the user cost converts to the price of acquiring R&D assets into a flow of capital services.

Samuels explained that linking the estimates of R&D embedded in value added from the BEA-BLS Industry-level Production Account to measures of value-added content embedded in gross exports using U.S. datasets that cover 1997–2019 gives the R&D share and the totals of R&D embedded in U.S. value-added exports and in GDP. It turns out that R&D content embedded in value-added exports is significantly higher than R&D content embedded in GDP. The GDP share of R&D hovered around 3 percent, with perhaps a small upward trend. The R&D share in value-added exports started the period at around 6 percent and ended the period around 8.5 percent. These estimates suggest that, in 2019, embedded R&D was about $149 billion out of a total of about $1.7 trillion in U.S. value-added exports, which is a higher share than the software and information technology (IT) capital services that are embedded in U.S. value-added exports.

Another finding from this approach, Samuels said, is that embedded R&D in value-added exports originate from a few industries. These industries include the chemical sector, the computer and electronic products industry (not surprisingly), and miscellaneous professional and scientific and technical services, which includes the more narrowly defined R&D sector itself. That is, it is these industries that conduct the majority of the R&D that gets embedded in gross exports throughout the entire economy.

Samuels believed the data he discussed are reasonably reliable but could be improved. One area for improvement involves the investment data that feed into BEA's estimates of capital. These data come from the NSF BERD (Business Enterprise R&D) survey, including estimates of spending on R&D, which BEA in turns adjusts to translate spending into investment. These estimates are by funder of R&D and by company, but BEA needs estimates by establishment and more industry detail. The price and depreciation of R&D investment (which feed into the estimates of the R&D capital stock and capital service measures) is an active area of research. Estimating a rate of return for R&D, which is used in the capital services calculation, is particularly difficult because R&D is inherently risky, and building this risk into the measurement of capital services is a challenge.

The second major dataset required for Samuels' approach is input-output data and, in particular, the domestic requirements table, which essentially takes total requirements and splits out imports. However, BEA lacks good data on which imports are used by which industry, so it employs an input proportionality assumption to spread imports across industries based on purchases.

Another major assumption underlying Samuels' approach is homogeneity of production within industries. Strong evidence suggests that industries that export have different production processes than those that focus on domestic markets. Consequently, one must assume that all establishments within the summary-level industries have the same production technology when one is applying the value-added and KLEMS splits at each chain of production, even though this appears to be at odds with the data.

Additional industry detail would help in identifying R&D's role in exports and would improve estimates of value-added trade overall. Specifically, it would be possible to define the interindustry purchases necessary to parse out what is coming from where and all the upstream effects that are important when taking the final step to estimate value-added content of exports. Samuels said that BEA has ongoing work in many of these areas. His assessment is that the prospects are significantly greater than the challenges.

Role of Government R&D

Thompson asked Samuels how government R&D is measured in estimating value-added exports. As a long-time agriculture economist, she knows that there is a lot more R&D in corn than that coming from the chemical industry. What about the contributions of the National Weather Service and other government programs? Samuels said that he chose farm as a potentially provocative example. The output of the government does not show up as purchases by an industry. There are good reasons why the official estimates handle government R&D that way, but Samuels agreed that it would benefit estimation of value-added R&D to adjust the official estimates so that government R&D is accounted for as an input. Samuels said there is a survey that attempts to measure R&D by the government sector, and those estimates exist in BEA's accounts, but they do not at present feed through to the input-output tables.

U.S. AND INTERNATIONAL WORK ON GLOBAL VALUE CHAINS: PAST AND FUTURE

Francisco Moris (NSF) served as the discussant for the session. His goal was to link the presentations to U.S. and international work in the area, identify measurement challenges and opportunities, and present principles for integrating the challenges for measurement of GVCs and intangibles into a "methodologic research agenda."

International Work on Measuring Globalization and Intangibles

Moris noted that globalization and intangibles have been a constant preoccupation for official statistics for a long time for surveys and national account purposes. This conference built on reports and conferences over the past 15 years on the intersection of globalization and intangibles and the challenges for official statistics (see Ahmad et al., 2020; Houseman and Mandel, 2015; Houseman and Ryder, 2010; NBER, 2006, 2018; Reinsdorff and Slaughter 2009; Upjohn, 2013; Upjohn and NAPA, 2009).

Moris said that GVCs and intangibles, particularly knowledge flows, have impacted the way everyone thinks about economic statistics. He cited national and international manuals that cover relevant topics:

- 2009 *System of National Accounts 2008* (SNA) treated R&D as investment and provided for national/international accounts (European Commission et al., 2009); BEA and NCSES developed an R&D Satellite Account in 2006 with estimates back to 1959 (Okubo et al., 2006, Section 1); 2013 BEA National Income and Product Accounts (NIPA) Comprehensive Revision incorporated R&D in U.S. GDP and related statistics (Kornfeld, 2013); other countries have done much the same.
- 2009 *Balance of Payments and International Investment Position Manual (BPM6)* provided guidance on treatment of R&D (see especially Table 10.4., Treatment of Intellectual Property, which includes R&D, software, and other intangibles) (IMF, 2009).
- 2010 *Handbook on Deriving Capital Measures of Intellectual Property Products* provided authoritative guidance (see especially Chapter 2, R&D; Chapter 4, Software/Databases) (OECD, 2010).
- 2010 *Manual on Statistics of International Trade in Services (MSITS)*, provided guidance on foreign affiliates statistics and R&D services (see Annex: Extended Balance of Payments Services [EBOPS] Classification) (United Nations et al., 2012).
- 2012 *Guide on Impact of Globalization on National Accounts* (see Chapter 7) (UNECE, 2011) and 2015 *Guide to Measuring Global Production* (see Chapter 4) (UNECE, 2015) provided guidance on intellectual property products (IPPs).
- 2015 *Organisation for Economic Co-operation and Development (OECD) Frascati Manual* provided authoritative guidance on R&D statistics (see especially the new R&D globalization chapter) (OECD, 2015).

Moris noted that all this work reflected the evolution in economic theory that previous workshop sessions discussed, such as thinking about economic growth in terms of endogenous growth, looking at increasing returns to trade and the role of trading in intermediate inputs, and thinking about MNEs, particularly how companies have moved from production fragmentation to R&D fragmentation. Statisticians and researchers from many countries worked together and eventually convinced official statisticians to be comfortable treating R&D as within the scope of GDP. This was a very important change internationally and for every country.

Session Papers in the GVC Data Space

Moris illustrated the GVC data space with a graphic (see Figure 14-3). The figure links official statistics (upper left) and big data (upper right) to macroaggregates, such as national accounts. It also links big data with product, market, and industry GVC case studies (lower right). Macroaggregates, in turn, generate challenges in terms of data gaps and opportunities in terms of new ways to measure GVCs and intangibles. Moris identified a missing link between case studies and macroaggregates (shown as a dotted line in the graphic). Moris located the presentations in the session in Figure 14-3 as follows: Lemmers from Statistics Netherlands fits with official statistics; Samuels from BEA fits with official statistics and macroaggregates; and Sturgeon noted the need for cross-field collaboration and compatibility of classification systems, which would facilitate cross-fertilization of macroaggregates and case studies.

Expanding on the session presentations, Moris observed that Lemmers provided a flavor of some of the methodological challenges in measuring global chains and national and international flows. Samuels described efforts by BEA and other agencies on linking, improving, and integrating data on intangibles for macroaggregates. NCSES just began a project with BEA on intangibles, including data needs and methodologies for supply use and for trade in value added (TiVA).

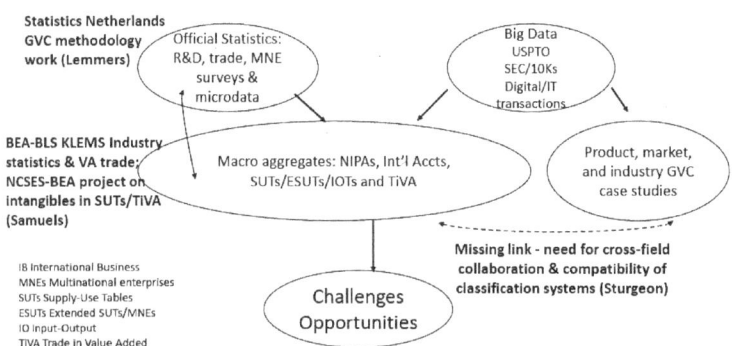

FIGURE 14-3 Session papers in the GVC-data space.
NOTE: BEA = Bureau of Economic Analysis; BLS = Bureau of Labor Statistics; GVC = global value chain; IT = information technology; KLEMS = K-capital, L-labor, E-energy, M-materials, and S-purchased services; NCSES = National Center for Science and Engineering Statistics; NIPA = national income and product accounts; R&D = research and development; SEC = U.S. Securities and Exchange Commission; USPTO = U.S. Patent and Trade Office; VA = value added.
SOURCE: Presentation by Francisco Moris.

Sturgeon has done relevant work on a variety of topics, but Moris wanted to emphasize his point about the need for better linking of data and the value of case studies.

Data Challenges and Opportunities

Moris noted that, even though the R&D definition is based on the Frascati and other manuals, there are still many practical issues for any survey or macroaggregate dealing with globalization and intangibles. He went on to list four such issues:

- **Organizational complexity**—As complex and important as MNEs themselves are, many global production arrangements involve contractors and R&D design companies, big and small, adding to the complexity to be measured. It is difficult enough to capture production activity among complex organizations across borders, let alone to capture activity involving intangibles (see UNECE, 2015).
- **IPP supply and use**—Another challenge is to distinguish production of knowledge—the supply of R&D, perhaps—from its use and ownership. For intangibles, this is difficult by definition because intangibles are often jointly produced and used within and across organizations. This phenomenon obscures who benefits from R&D or knowledge in general and where (see IMF, 2009; UNECE, 2015).
- **Classification issues**—One issue is the focus of transactions that a given survey asks about. Is it current-year R&D? IPP charges from past R&D? R&D commingled with other intangibles? This gets complicated for classification of products or services for balance-of-payment purposes or classification systems for R&D and industry, such as companies that do manufacturing only under contract or companies that provide design services. Another classification issue is the distinction between property income and payment for services (see Chapter 7, "International Transactions in Intellectual Property Products," UNECE, 2011).
- **Valuation**—One question is how close is a reported value to an arm's-length transaction, and therefore how can it best be priced (see OECD, 2017).

Opportunities for Collaboration

Moris identified opportunities for intracountry collaboration by national statistical agencies with administrative agencies and big data sources, including survey development, concordance of business/MNE registers, large-scale data-linking and integrated business statistics programs, and aggregates for IPPs in national and international accounts. Opportunities for international collaboration (bilateral, regional, within international organizations) include implementing IPP material in the manuals for statistics on globalization cited previously, developing a three-way concordance among MSITS/EBOPS 2010, CPC (central product classification), and the new Eurostat Business Functions platform, and establishing working groups on a research agenda for the SNA.

Principles for Data Integration

Moris finished by offering a couple of principles for integrating research on methods for measuring GVCs and intangibles. Most important, future work should build on international research agendas and collaborations that are already in place and then move from there.

He proposed an integrative framework to summarize challenges, possible methods, and implementation issues. The framework has three domains: R&D/science, technology, and innovation (STI) indicators; services trade/GVCs; and SNA/NIPA aggregates. It also has three issue areas: challenges and research questions, possible methods, and implementation issues. Entries in this matrix would identify projects in terms of high, medium, and low priority (see Figure 14-4).

ADDITIONAL COMMENTS

Thompson thanked everyone for keeping so many complex ideas in front of the workshop participants. She began an open discussion period with a comment. When she started work at BEA in the international area, she wanted

Issues: priorities/timeframe	Domains: R&D/STI indicators (e.g. MNE IPPs)	Services trade/ GVCs (outsourcing, intra-MNE flows)	SNA/NIPA aggregates (SUTs/TiVA)
* Challenges and Research Questions			
high-medium-low			
* Possible Methods			
high-medium-low			
* Implementation issues			
near-medium-long term feasibility			

FIGURE 14-4 Principles and integrative framework—Methodological research agenda on GVCs and intangibles.
NOTES: GVC = global value chain; IPP = intellectual property product; MNE = multinational enterprise; NIPA = national income and product accounts; R&D = research and development; SNA = System of National Accounts; STI = science, technology, and innovation; SUT = supply-use table; TiVA = trade in value added.
SOURCE: Presentation by Francisco Moris.

BEA researchers to understand the complexity of measuring GVCs. She invited an executive from a large computer services corporation to speak about the company's intellectual property and how they managed the billing for that within their company and how the work was organized.

She learned that the company did it all through their people's salaries in different parts of the world. With staff in Brazil, Germany, and India, they were creating new high-tech software for customers, whom they billed from headquarters. It is hard to think about how to apportion these costs among the various countries' economies. She said the large case unit approach taken by some national statistical agencies makes a lot of sense, particularly for very big companies.

Sturgeon said he was struck by Lemmers' remarks about how to fill in some of the data gaps around R&D by profiling multinational firms or firms using GVCs to source R&D. Instead of talking to an important IT services firm in one's country, the task is now to make a global odyssey to talk with big companies to find out how they manage production and R&D. Globalization has made everyone's life more challenging in terms of research.

Samuels wanted to call out the collaborative project Moris mentioned between BEA and NCSES to use the official data to refine the estimates of value-added chains and the role of the United States, using more industry detail and looking at the role of intangibles. Samuels is looking forward to seeing public results from that project in the near term.

Moris said that, given the many colleagues from overseas in the conference, he wanted to emphasize that the United States has done a great deal of work with colleagues at OECD and other organizations. For an issue so complex as globalization and intangibles, international collaboration is fundamental for countries to learn from each other how to better understand the effects of intangibles on productivity and output.

Lemmers added his support for talking to firms to learn how they work, and adapting surveys accordingly and testing concepts. Such testing is essential for determining practical measures. If testing shows that a concept is not working, then statistical agencies need to change course.

15

Lessons from the Workshop: A Panel Discussion

Susan Helper (Frank Tracy Carlton professor of economics, Weatherhead School of Management, Case Western Reserve University, and workshop cochair) said she was pleased to introduce the closing panel of this workshop on innovation in global value chains (GVCs)—Rebecca Riley (professor of practice in economics, King's Business School, King's College London), William Powers (chief economist and director of the Office of Economics, U.S. International Trade Commission),[1] Maria Borga (deputy division chief, Balance of Payments Division, International Monetary Fund), and Nadim Ahmad (deputy director, Centre for Entrepreneurship, SMEs, Regions and Cities, Organisation for Economic Co-operation and Development [OECD]). Each panelist was asked to think about key takeaways from the workshop (see Box 15-1).

PERSPECTIVE FROM THE UNITED KINGDOM

Riley noted the complexities of the challenges facing national statistics agencies in producing the information needed by decision makers for understanding GVCs. She made points on policy needs, data sources and access, and collaboration.

Policy Needs

Riley emphasized the importance and demand for the types of analytical outputs produced and discussed in this workshop for informing policy. The

---

[1] Powers' views expressed here are solely his. His comments in this chapter are not meant to represent in any way the views of the International Trade Commission or any of its individual commissioners.

> **BOX 15-1**
> **Selected Workshop Takeaways from Panel Members**
>
> - Policymakers want statistics and analysis on global value chains (GVCs) and their effects on trade, value added in different industries, and within-country economic growth or decline—the "geographies of discontent."
> - Data confidentiality is important, but innovative ways of sharing business data need to be found, including commercial data, for national statistics and research—the public good argument for sharing is strong.
> - Much more granular detail is needed on firm ownership and location, particularly for multinational enterprises (MNEs).
> - Microdata linked across different domains are needed (in a secure environment) to study GVCs and MNEs.

outputs are not necessarily what the United Kingdom calls national statistics, which have a special status and are produced according to a code of practice. The United Kingdom would describe the outputs as experimental and investigative. She provided several examples.

As Riley pointed out, the United Kingdom is a relatively small and open economy compared with the United States. A lot of its research and development (R&D) is undertaken by foreign firms, and one of its biggest services exports is education. The United Kingdom has also just parted from its biggest and nearest trading partner following the vote to leave the European Union in 2016, which will have implications for labor mobility and trade frictions. In this context, it is hard to overestimate the importance of the trade and value-added data produced by the OECD. But there is, of course, always a demand for more detail. The United Kingdom has used detailed customs data for this purpose, despite the potential flaws in this approach. She and her collaborators were able to produce estimates of direct domestic value added in exports by detailed manufacturing sectors. This is very hard to do for services, but they also looked at financial services with data from the Bank of England.

Basically, Riley said, she and her collaborators produced an important analytical output product. These microdata are not national statistics, but they do provide information on the value added of exports in specific manufacturing industries (~200 groupings) and where jobs are likely to be affected by changing trade patterns. So policy makers had an indication, not only as to the dependency of particular industries on imports in their production processes, but also where there might be economic impact.

Policy makers also want to understand the local effects of changing trade arrangements in the United Kingdom and the implications of these changes in value added for different parts of the country. Riley and her collaborators looked at this with colleagues in Groningen, basically attempting to map estimates by industry back onto the country, but the results were mixed. A key weakness is the lack of data, not only on networks internationally, but also on domestic networks because the fragmentation in production is taking place at many levels. Their recommendation was to create better supply-use tables among the different areas of the United Kingdom.

According to Riley, another area of intense policy interest is around productivity growth, or increasingly, its lack of growth. The slowdown is very unevenly spread, and it has huge implications for living standards across the United Kingdom. Interest in these issues has led to a significant investment in a productivity institute and also in John Van Reenen's innovation and diffusion program to help address these productivity puzzles specifically. Riley was very interested to see Van Reenen's findings using Belgian data on diffusion of good practice or innovation to suppliers of superstar companies (see Amiti et al., Chapter 8), and also Surana's paper on the importance of innovation among affiliated small- and medium-sized enterprises (Surana et al., Chapter 6). Networks of innovation need to be better understood to address the productivity puzzle.

Data Sources and Access

Riley commented on the huge amount of data work that went into several of the papers in the workshop. There are many novel applications, and there has been investment in developing extensive microdata. There are also examples of looking to alternative data sources to inform what is going on in GVCs.

Riley wanted to highlight some of the potential uses of new data sources (e.g., web scraping) for illustrating the connectivity among companies and understanding their spatial allocation, and for understanding the extent of factoryless manufacturing in the United Kingdom, such as in work by Coyle and Nguyen (2020).[2] Unlike the United States, the United Kingdom does not have survey data on factoryless manufacturing, and Riley would like to see questions on the topic added to surveys, given the growth in this phenomenon.

Riley stressed the need for effective data access and sharing mechanisms. The United Kingdom passed the Digital Economy Act in 2017, with the idea that it would facilitate sharing of microdatabases across government departments and with academics. A recent change in legislation has made it easier to access trade data because of policy changes resulting from the break with the European Union. Riley said these changes in legislation have been useful but more dedicated resources are needed, given that the investments required for data linkage and analysis are significant.

Collaboration

Riley highlighted the need for collaboration between academics and statistical agencies to produce useful analytic outputs, which in turn requires data sharing. Many of the insights the United Kingdom has been able to share about

---

[2] Factoryless manufacturing is when a company contracts with another firm for manufacturing.

trade in the current environment have only been possible because of data sharing across statistical agencies.

PERSPECTIVE FROM AN INTERNATIONAL TRADE ECONOMIST

William Powers indicated that, while his views are his own and he is not acting in an official capacity, his remarks are informed by interactions with stakeholders at the International Trade Commission (ITC). The ITC does not make policy, but when requested, it provides trade policy advice and support to both the U.S. Congress and the executive branch.

Powers said the ITC has seen a big focus on GVCs and firm strategies within them, and the papers at the workshop have commendably sought to dig closely into those strategies and motivations. Powers appreciated that the organizers and panelists focused on new work, which led to a robust discussion of methodology and data sources for which perhaps there is no clear consensus yet as to the right approaches to measurement. Powers thought he could usefully talk about areas worthy of more exploration.

First, Powers said that academics tend to think of the effects of policy on firms and the location of innovation, but it is equally important to think about how firms, at least in the aggregate, affect policy. To give examples, consider the U.S.–China Phase 1 deal nailed down in 2020. The portions that generated a lot of press play and media attention were agriculture and expansion of trade. These provisions were actually prompted by intellectual property and technology transfer issues. If one looks at the agreement itself—it is relatively short, about 60 pages—quite a lot of text is devoted to intellectual property.

Another example Powers gave is the American Jobs Plan that came out about in April 2021 from the Biden administration, which focused on much of what the workshop has discussed: revitalizing manufacturing, securing supply chains, investing in R&D, and training Americans for jobs. The plan has a definite goal to focus on both innovation and manufacturing and production within the United States. Powers asserted that it was important to communicate insights from the workshop to policy makers, just as Riley suggested connecting with national statistical agencies. Powers has seen that policy makers and other stakeholders have a real appetite right now for empirical analysis—certainly, on trade policy and likely for supply chains for innovation policy. If analysis is accessible, they are interested in it, and they are asking for it, so researchers should think about communication with policy makers as a venue for getting one's work out there.

Lastly, if an informed group, such as the participants in this workshop, does not share their work, Powers stated that other people would. He said it was important to disseminate solid analysis. The issue of restoring American competitiveness has been around since at least the 2008–2009 global downturn. The analyses presented at this workshop are providing some real answers.

Powers noted the clear documentation of retrenchment and plateauing of international GVCs after 2011 and further retrenchment since 2014, despite the rise at the same time of domestic GVCs—that is, domestic outsourcing and not offshoring (see Wang et al., Chapter 3). Powers said his own work clearly shows retrenchment in internationalization of GVCs. He wondered if the same phenomenon is also happening in innovation and whether the two are connected and worth considering.

PERSPECTIVE FROM AN INTERNATIONAL FINANCIAL ORGANIZATION

Maria Borga said she would like to focus—as a main takeaway—on the update of international statistical standards. Her organization, the International Monetary Fund (IMF), launched a process for updating both the System of National Accounts (SNA) and the balance of payments in 2020. At this stage there are task teams working on developing recommendations, organized around such themes as globalization or digitalization, as well as for specific parts of the accounts, such as the current account or direct investment. These task teams include representatives of countries and international organizations. Some have proposals that are being discussed, while other teams need to do more work.

The first thing that struck Borga was the need for more information on multinational enterprises (MNEs). This need, fortunately, is well recognized by most statistical agencies and countries, and recommendations have been put forward that are on track. One recommendation is for the SNA to include, in the institutional-sector accounts, additional breakdowns of financial and nonfinancial corporations among foreign-controlled firms, domestic MNEs, and other domestic firms. Borga observed that there has not been support for the whole extended supply-use tables framework because it was considered too burdensome, but there is no reason that countries cannot continue to do this work, given its extraordinary payoff.

Borga wanted to highlight a proposal for the balance of payments—namely, presentation of the balance of payments current account by enterprise characteristics. It would take the trade in goods measured on a balance-of-payments basis and break it down between the categories of foreign-controlled, domestic MNE, and other domestic enterprise, as well as by firm size. It would also provide additional details on the top trading partners, industries, and products. There would be a similar breakdown, not only for services, but also for receipts and payments of investment income. Firm characteristics would also be available. Borga believes this proposal responds to some of the issues raised in the workshop, such as implicit payments, especially for things like intellectual property that happen through direct investment income rather than through explicit service payments.

Borga said her second takeaway was related to MNEs—namely, looking at the nationality of firms not only in terms of ownership but also in terms of location. Fundamental to the way accounts are done is residency, domestic or foreign. Additional information on nationality or ownership could be very helpful.

Borga talked about a specific proposal for compilation of data for foreign direct investment (FDI) statistics, which are close to her heart. The statistics reflect when there are direct links between firms in two countries; but when there are chains of ownership, there need to be supplemental statistics on the ultimate investing economy and the ultimate host economy. In developing these statistics, one issue often unrecognized is the need for consolidation. She was surprised by a comment by James Hines (University of Michigan, the discussant for Accoto et al., Chapter 4) about people not realizing that the income from tax havens could reflect income from further down the chain, because that is so well-known among the FDI statistics community. The FDI statistics proposal includes a way to identify the direct investment assets and liabilities in the host country versus those held further down the chain, and similarly for income payments generated in the host economy versus income that was just passing through that economy. These kinds of statistics would address the double counting issue.

Borga noted that people often believe it to be simple to identify the nationality of MNEs. In her view, such identification is becoming more and more difficult. The prominent example is something like a corporate inversion, in which a U.S. company, for example, will end up moving its headquarters to Ireland. Is it now suddenly an Irish MNE or is it still an American MNE? The IMF is working on that issue.

Borga said her last key takeaway had to do with the location of intellectual property, or intangible capital (see, e.g., Accoto et al., Chapter 4). Even when tax avoidance is not extensive, there is evidence of a large amount of intangible capital residing in tax havens. This phenomenon raises the question of whether there is really economic ownership of that intangible capital or just tax convenience.

Borga said there have been proposals to address this problem. For example, some say the intangible capital should be attributed to the parent firm because that firm ultimately owns the intellectual property, bears the risks, and reaps the rewards. Others say the intangible capital should be credited to where it was produced and no cognizance taken of its being in a tax haven. There are conceptual issues with this approach, which would treat intangible capital differently from physical capital, and there are practical issues as well. However, the workshop discussions have made clear that the increasing globalization of R&D calls into question the attribution to the parent, and cross-border collaboration also calls into question where intangible capital is developed and how it is assigned. More work is needed in this area, according to Borga.

Borga concluded by giving a nod of approval to the paper on customs records (Krizan et al., Chapter 11). That work should be very helpful to the IMF.

PERSPECTIVE FROM OECD

Nadim Ahmad said that a key takeaway for him was the need for more quantity and granularity of data to tackle many of the issues involved in understanding globalization. There is also enormous potential to capitalize on existing data and bring them to bear. He believes it is important to build on these types of initiatives to mainstream them into official statistics and analyses. Even if that is not achievable, mechanisms are needed to prevent others having to reinvent the initiatives by workshop presenters. These initiatives should be turned into long-lived assets. They themselves become R&D and sources of innovation from which others can draw.

Ahmad reported on efforts under way internationally. OECD, for probably two decades now, has been trying to obtain microdata, which are the only means to answer many questions about globalization. There are challenges, of course. Official microdata are only available under severe confidentiality restrictions. But there has been a lot of progress over the last 10–20 years, driven by events such as this conference, coupled with the realization that existing data could be used in a much better way. There is now momentum within OECD to develop a microdata laboratory. The idea is that countries would give OECD a whole range of microdata that could be stored safely in the laboratory. This asset would allow OECD to link microdata across different domains to study, for example, innovation within trade, or trade in the national accounts, and of course the MNE perspective.

Ahmad stated that OECD wants to go yet further to tackle issues not yet studied. Ahmad echoed a point Riley made about the critical importance of location for understanding trade and the backlash to trade in some regions in the past few years and the retrenchment of globalization as a consequence. OECD refers to "geographies of discontent" for this phenomenon. Ahmad said one of the principal needs is to build location into national accounts. To be serious about understanding how trade manifests itself through systems and how it impacts people requires a granular place-based approach. Business registers within countries are place-based, and some countries have firm-to-firm identifiers. It is possible to use value-added tax (VAT) records, for example, to match transactions in the system, which in theory would enable one to create very detailed supply-use tables.

With that information, Ahmad continued, a lot of the questions generated by workshop discussions would be much easier to answer, including the ability to adopt a firm-to-firm perspective and distinguish between MNEs and non-MNEs and large and smaller firms. Ahmad stressed that bringing microdata into the system has to be a primary goal to allow the granularity needed for policy work. As Sturgeon has noted on many occasions, the granularity within conventional estimates for trade in value added (TiVA) provides a nice view of what is going on but is insufficient for much policy analysis.

Ahmad's last point, in terms of painting a roadmap for what is needed, is to recognize the growing momentum behind responsible business conduct and corporate social responsibility. These things are going to be increasingly important as global supply chains become more driven to ensure that suppliers are abiding by corporate social responsibility rules.

Ahmad thought it could be useful to embody responsible business conduct in official statistics, in order to motivate its uptake by all firms. At least, there could be some type of international mechanism allowing every firm to report how it is upholding responsible business chains. Ahmad also said it is important to understand how responsible business conduct impacts productivity—the notion, for example, that corporate social responsibility can be a way of improving and introducing efficiencies in the production process.

DISCUSSION

Helper posed a question, noting the variety of datasets and countries represented in the workshop. She asked if the panelists thought there was a particular country's approach or data series within the collective of financial statistical agencies that seemed promising. Her particular interest in constructing GVCs is understanding the flows from firm to firm. Input-output tables delineate the flows from industry to industry, but it is much harder to see firm-to-firm flows. This dynamic can be seen to some extent in the international trade data in the United States, when products cross country borders, but there is nothing comparable for flows within the United States. Countries that have VAT systems can generate that data.

Ahmad did not want to single out particular countries, but he thought that many of the Scandinavian countries exemplified good practice because of their general openness with data and recording of information in their statistical systems. In fact, many Nordic countries provide considerable amounts of publicly accessible data, such as individual salaries. Such data facilitate linked datasets, making it much quicker to proceed from a research idea to a new statistic, compared with the 20 years cited by Sturgeon. Two examples are estimates of TiVA and trader/buyer characteristics.

Ahmad stated that national statistical organizations recognize that they ought to be much quicker to deliver needed information. Because of powerful systems that allow tapping into and integrating relevant data, they are able to deliver in a way they were not before.

In terms of countries that are producing relevant data on the firm-to-firm level, Ahmad said, Costa Rica, Chile, and Belgium deserve plaudits because they have been able to link relevant information in a very powerful way. Lemmers, who is now chair of an OECD expert group on extended supply-use tables, is also looking at firm-to-firm measurement.

Powers said that Fort pioneered using information in the U.S. system to analyze domestic outsourcing as opposed to foreign offshoring (Fort et al., Chapter 13). There is no public dataset to support exploration along these lines. Powers has been disappointed not to see more work of this kind.

Borga supported the idea of case studies of big firms, especially when it comes to understanding something like FDI. The harder step with which the field is still struggling is cross-country comparisons and enabling statistical agencies to compare what a company tells country A versus country B. There are some global initiatives trying to profile MNEs in an open way so people can have access to the same information about the company.

Riley completely agreed that cross-country comparability is crucial, so the microdata laboratory at OECD sounds very interesting. There is huge potential in firm-level data and in linking with data on employees and all sorts of surveys. Orbis data on companies around the globe are used in many of these types of studies, but Orbis has a lot of missing data partly because the data are not designed to be used by researchers.[3]

If everyone is using these data and developing code to be able to use them in comparable ways, there could be a case for having a code-sharing facility that is quite broad in its nature. Riley wondered if OECD could consider compiling some of these international commercial databases themselves so that their value as a public good could be fully realized.

Ahmad responded that OECD had thought about this idea. The difficulty is the proprietary nature of the datasets, which makes that a real challenge. Even if OECD, World Bank, IMF, and other organizations were to put together an international consortium to acquire the data to be a common resource, the restrictions that each organization has on using the data means that sharing would be a challenge.

Ahmad continued that the other challenge is dissemination. OECD and other international organizations are allowed to use these data for their analyses, but it is difficult to disseminate firm-level information back to users. This situation is exactly why a microdata laboratory, which used public information collected as a public good by national statistical offices, would be a better approach.

Ahmad said that OECD's goal is to bring government and commercial information together within OECD and make it available, certainly in the first instance to all OECD staff. There are real hurdles to overcome in terms of getting broader access to microdata. OECD hopes it will be able to prove that it can be a trusted partner in terms of making and keeping private-sector data safe and allowing others in the public sector to access data in the laboratory. That is basically the plan.

Ahmad shared that one way OECD is trying to motivate the idea of a microdata laboratory is to allow national statistical organizations themselves to see that they, too, could benefit from this process. A very simple benefit would be reconciliation of trade data, with the availability of firm identities supporting data linkage of exports by one firm and imports by another firm. Ahmad said OECD at present is trying to balance international trade in goods and services using available data, which comes with a certain degree of aggregation, which means that there are some assumptions and some shortcuts. OECD believes it is doing the best that can be done, but could do better if there were access to microdata.

Helper relayed two questions from workshop participants, one quite detailed and one quite broad. The first question was about special purpose entities (SPEs). Some artificial subsidiaries inside countries act similar to SPEs across borders of states or provinces. The SNA mandates consolidation of these at the national level. How does that affect research on innovation of GVCs? The second question was about (1) how to handle the challenge of protecting confidentiality;

---

[3] See Orbis, Compare Private Company Data, Bureau van Dijk (https://www.bvdinfo.com/en-gb/our-products/data/international/orbis).

and (2) how to handle the loss of identification of most transactions resulting from the organization of firms into GVCs.

Ahmad agreed that confidentiality is a critical issue. OECD hopes to be able to acquire data under circumstances that allow for data sharing and reuse. Also, Ahmad agreed that the way SPEs are defined in the system creates difficulties. If the SPE is a resident unit, the activities recorded there will show up in GDP. But some countries, such as the Netherlands, provide estimates without SPEs, and Ahmad believes that is a good approach.

Ahmad wanted to respond to a question from Sturgeon about the utility of a global register. Ahmad agreed completely that having a global register would help address a number of questions—for example, the importance of MNEs, which is currently overlooked. Right now, statistics are developed around national accounts, but increasingly there is a need to recognize the international concept and to ascertain the footprint of the MNE.

Ahmad explained how OECD was using web-scraping tools to develop a database called ADIMA (the analytical database of individual multinationals and their affiliates), which so far has profiles for 500 MNEs. OECD hopes this database will be the launch pad for a global register at some stage.

Returning to the question of confidentiality, Borga spoke from her experience at a national statistical agency, saying that it is very important and is key to getting firms to provide information. She agreed, however, that there is a need to modernize views of confidentiality. A few years ago, Statistics Canada changed how it approached confidentiality, resulting in the release of much more granular detail. Borga said confidential information is shared for tax purposes. It would be great if there was recognition that sharing business data among statistical agencies could be done in a way that would not harm the companies.

Borga said a global register and more information on very large multinationals would be very helpful to statistical agencies. Work in this area has been under way for a long time, but progress has been accelerating.

Powers wanted to respond to the comment that because of the organization of GVCs, identification of most transactions is lost. Powers said the challenge is even tougher because of acquisitions of firms in one country by domestic companies in another country. It would be difficult to capture these kinds of organizational relationships in a global register.

Powers said the good news, at least in the United States, is the uptick in the appetite for empirical information and using that information to make policy. Correspondingly, there is recognition that data gaps and data siloing make providing timely input to policy makers more difficult.

Riley said she agreed that confidentiality is important but would also argue that the case for public good is often quite strong. When arrangements are in place in which researchers have absolutely no incentive to breach confidentiality rules, it is perfectly compatible to use data for research for the purpose of public good and to maintain confidentiality. As an example, Riley pointed to a huge increase in data sharing during the pandemic because there was suddenly a very strong imperative to get needed information and to relate databases in order to understand what was happening. There is a very good case for public good in that instance. Riley expressed hope that these types of sharing arrangements can be maintained going forward. She also supported a global register.

CLOSING REMARKS

Wolfgang Keller (University of Colorado Boulder and workshop cochair) observed in closing the workshop that, when thinking about globalization, certain waves are often distinguished. There was the emergence of large-scale international trade in the late 19th century; then FDI starting in the 1970s; and, most recently, complex production networks, GVCs, and innovation networks emerging around the year 2000. It is easy to overstate the implications of the most recent developments. For example, as early as the 19th century, Jardine Matheson—a British trading company out of Hong Kong that was founded in the 1830s—already had complex networks with vertical specialization connecting Europe, British India, China, and Japan. On the other hand, it is clear from the papers at the workshop that the rising importance of intangibles and the knowledge economy gives rise to important new questions.

Keller stated that one of the jobs of economists is to provide to the world's central bank chiefs estimates of the rate of innovation or productivity growth so they can target the correct level of money supply. To take another example, the world's treasury secretaries need input on what should be the rate of taxation on global companies. Key to getting these numbers right is relying on good data and measurement, which is why investing in statistical agencies and fostering exchanges with researchers are important. Keller hoped the workshop contributed to this needed effort.

References

Acemoglu, D., and P. Azar. 2020. Endogenous production networks. *Econometrica* 88(1):33-82.

Acemoglu, D., P. Aghion, L. Bursztyn, and D. Hemous. 2012. The environment and directed technical change. *American Economic Review* 102(1):131-166.

Acemoglu, D., A. Kakhbod, and A. Ozdaglar. 2017. Competition in electricity markets with renewable energy sources. *The Energy Journal* 38. KAPSARC Special Issue.

Adão, R., C. Arkolakis, and S. Ganapati. 2020. *Aggregate implications of firm heterogeneity: A nonparametric analysis of monopolistic competition trade models*. Working Paper w28081. Cambridge, MA: National Bureau of Economic Research.

Aghion, P., A. Bergeaud, and J. Van Reenen. 2021. *The impact of regulation on innovation*. Working Paper w28381. Cambridge, MA: National Bureau of Economic Research.

Ahmad, N., B. Moulton, J. D. Richardson, and P. van de Ven, eds. 2020. *The challenges of globalization in the measurement of national accounts*. Preliminary Drafts. Chicago, IL: University of Chicago Press. https://www.nber.org/books-and-chapters/challenges-globalization-measurement-national-accounts.

Aitken, B., and A. Harrison. 1999. Do domestic firms benefit from direct foreign investment? Evidence from Venezuela. *American Economic Review* 89(3):605-618.

Alfaro-Urena, A., I. Manelici, and J. Vasquez. 2019. *The effects of multinationals on workers: Evidence from Costa Rica*. Working Paper. Private Enterprise Development in Low-Income Countries. https://pedl.cepr.org/publications/effects-multinationals-workers-evidence-costa-rica.

Alfaro-Urena, A., I. Manelici, and J. Vasquez. 2021. *The effects of multinationals on workers: Evidence from Costa Rican microdata*. Working Paper. Princeton, NJ: Princeton University, Department of Economics, Center for Economic Policy Studies. https://EconPapers.repec.org/RePEc:pri:cepsud:285.

Alstadsæter, A., N. Johannesen, and G. Zucman. 2018. Who owns the wealth in tax havens? Macro evidence and implications for global inequality. *Journal of Public Economics* 162:89-100.

Alvarez, R., and R. López. 2008. Is exporting a source of productivity spillovers? *Review of World Economics* 144(4):723-749.

Antràs, P. 2003. Firms, contracts, and trade structure. *Quarterly Journal of Economics* 118(4): 1375-1418.

Antràs, P. 2014. Grossman–Hart (1986) goes global: Incomplete contracts, property rights, and the international organization of production. *Journal of Law, Economics, and Organization* 30(suppl_1):i118-i175.

Antràs, P., and D. Chor. 2013. Organizing the global value chain. *Econometrica* 81(6):2127-2204.

Antràs, P., and D. Chor. 2021. Global value chains. In *Handbook of International Economics*, vol. 5. Amsterdam: Elsevier.

Antràs, P., and E. Helpman. 2004. Global sourcing. *Journal of Political Economy* 112(3):552-580.

Antràs, P, and E. Helpman. 2009. Contractual frictions and global sourcing. In *The organization of firms in a global economy*, edited by E. Helpman, T. Verdier, and D. Marin. Cambridge, MA: Harvard University Press. Pp. 9-54.

Antràs, P., T. Fort, and F. Tintelnot. 2017. The margins of global sourcing: Theory and evidence from US firms. *American Economic Review* 107(9):2514-2564.

Aoki, K., and M. Wilhelm. 2017. The role of ambidexterity in managing buyer–supplier relationships: The Toyota case. *Organization Science* 28(6):1080-1097.

Argyres, N. 2013. Contracting for innovation. In *Innovation and growth: What do we know?*, edited by A. Thakor. Singapore: World Scientific.

Arkolakis, C., N. Ramondo, A. Rodríguez-Clare, and S. Yeaple. 2018. Innovation and production in the global economy. *American Economic Review* 108(8):2128-2173.

Arto, I., E. Dietzenbacher, and J. M. Rueda-Cantuche. 2019. *Measuring bilateral trade in terms of value added*. Luxembourg: Publications Office of the European Union.

Atalay, E. 2017. How important are sectoral shocks?. *American Economic Journal: Macroeconomics* 9(4):254-280.

Atalay, E., A. Hortaçsu, J. Roberts, and C. Syverson. 2011. Network structure of production. *Proceedings of the National Academy of Sciences* 108(13):5199-5202. https://doi.org/10.1073/pnas.1015564108.

Atalay, M., N. Anafarta, and F. Sarvan. 2013. The relationship between innovation and firm performance: An empirical evidence from Turkish automotive supplier industry. *Procedia Social and Behavioral Sciences* 75:226-235.

Autor, D., D. Dorn, L. Katz, C. Patterson, and J. Van Reenen. 2017. *The fall of the labor share and the rise of the superstar firms.* Working Paper w23396. Cambridge, MA: National Bureau of Economic Research. https://www.nber.org/system/files/working_papers/w23396/w23396.pdf.

Autor, D., D. Dorn, L. Katz, C. Patterson, and J. Van Reenen. 2020. The fall of the labor share and the rise of superstar firms. *Quarterly Journal of Economics* 135(2):645-709.

Bajgar, M., G. Berlingieri, S. Calligaris, C. Criscuolo, and J. Timmis. 2019. Industry concentration in Europe and North America. Discussion Paper 1654. London: Centre for Economic Performance. https://cep.lse.ac.uk/pubs/download/dp1654.pdf.

Barresse, G., F. Kamal, J. Miranda, and W. Ouyang. 2017. *Business dynamics of U.S. exporters: Integrating trade transactions data with business administrative data.* mimeo.

Barrios, S., and D. d'Andria. 2020. Profit shifting and industrial heterogeneity. *CESifo Economic Studies* 66(2):134-156.

Barro, R. J. 2021. Double counting of investment. *Economic Journal* 131(638):2333-2356.

Beer, S., and J. Loeprick. 2015. Profit shifting: Drivers of transfer (mis)pricing and the potential of countermeasures. *International Tax and Public Finance* 22(3):426-451.

Benguria, F. 2021. The matching and sorting of exporting and importing firms: Theory and evidence. *Journal of International Economics* 131.

Bernard, A. B., and S. Dhingra. 2015. *Contracting and the division of the gains from trade.* Working Paper w21691. Cambridge, MA: National Bureau of Economic Research.

Bernard, A., and A. Moxnes. 2018. Networks and trade. *Annual Review of Economics* 10:65-85.

Bernard, A. B., J. Eaton, J. B. Jensen, and S. Kortum. 2003. Plants and productivity in international trade. *American Economic Review* 93(4):1268-1290.

Bernard, A. B., J. B. Jensen, and P. Schott. 2006. *Transfer pricing by US-based Multinational firms.* Working Paper 12493. National Bureau of Economic Research. https://www.nber.org/papers/w12493.

Bernard, A. B., J. B. Jensen, S. Redding, P. Schott. 2009. The margins of US trade. *American Economic Review* 99(2):487-493.

Bernard, A. B., V. Smeets, and F. Warzynski. 2017. Rethinking deindustrialization. *Economic Policy* 32(89):5-38.

Bernard, A. B., J. B. Jensen, S. Redding, P. Schott. 2018a. Global firms. *Journal of Economic Literature* 56(2):565-619.

Bernard, A. B., A. Moxnes, and K. H. Ulltveit-Moe. 2018b. Two-sided heterogeneity and trade. *Review of Economics and Statistics* 100(3):424-439.

Bernard, A., E. Bøler, and S. Dhingra. 2018c. *Firm-to-firm connections in Colombian imports.* New York: Routledge.

Bernard, A., E. Blanchard, I. Van Beveren, and H. Vandenbussche. 2019. Carry-along trade. *Review of Economic Studies* 86(2):526-563.

Bernard, A., T. Fort, V. Smeets, and F. Warzynski. 2020. Heterogeneous globalization: Offshoring and reorganization. Working Paper 26854. Cambridge, MA: National Bureau of Economic Research.

Bernard, A., T. Fort, V. Smeets, and F. Warzynski. 2021. Heterogeneous globalization: Offshoring and reorganization. Working Paper 26854. Cambridge, MA: National Bureau of Economic Research.

Berry, H. 2014. Global integration and innovation: Multicountry knowledge generation within MNCs. *Strategic Management Journal* 35(6):869-890.

Bilicka, K. 2019. Comparing UK tax returns of foreign multinationals to matched domestic firms. *American Economic Review* 109(8):2921-2953.

Bilir, L. K. 2014. Patent laws, product life-cycle lengths, and multinational activity. *American Economic Review* 104(7):1979-2013.

Bilir, L., and E. Morales. 2020. Innovation in the global firm. *Journal of Political Economy* 128(4):1566-1625.

Binz, C., and B. Truffer. 2017. Global innovation systems—A conceptual framework for innovation dynamics in transnational contexts. *Research Policy* 46(7):1284-1298.

Bloom, N., C. Genakos, R. Sadun, and J. Van Reenan. 2012. Management practices across firms and countries. *Academy of Management Perspectives* 26(1):12-33.

Bloom, N., J. Van Reenen, and S. Melvin. 2013. Gokaldas exports (A): The challenge of change. Case SM213A. Stanford, CA: Stanford Graduate School of Business.

Bloom, N., R. Sadun, and J. Van Reenen. 2016. *Management as a technology?*. Working Paper w22327. Cambridge, MA: National Bureau of Economic Research.

Bloom, N., J. Van Reenen, and H. Williams. 2019. A toolkit of policies to promote innovation. *Journal of Economic Perspectives* 33(3):163-184.

Bloom, N., C. I. Jones, J. Van Reenen, and M. Webb. 2020. Are ideas getting harder to find?. *American Economic Review* 110(4):1104-1144.

Blum, B., S. Claro, and I. Horstmann. 2010. Facts and figures on intermediated trade. *American Economic Review* 100(2):419-423.

Blum, B. S., S. Claro, and I. J. Horstmann. 2018. Trade costs and the role of international trade intermediaries. *Handbook of International Trade and Transportation*. Cheltenham, UK: Edward Elgar Publishing.

Borin, A., and M. Mancini. 2019. *Measuring what matters in global value chains and value-added trade*. Policy Research Working Paper 8804. Washington, DC: World Bank.

Branstetter, L. G., B. Glennon, and J. B. Jensen. 2019. The IT revolution and the globalization of R&D. *Innovation Policy and the Economy* 19(1):1-37.

Brugues, F. 2020. *Take the goods and run: Contracting frictions and market power in supply chains*. Providence, RI: Brown University.

Bruner, J., D. Rassier, and K. Ruhl. 2018. Multinational profit shifting and measures throughout economic accounts. In *The challenges of globalization in the measurement of national accounts*. Chicago, IL: University of Chicago Press.

Cajal-Grossi, J., R., Macchiavello, and G. Noguera. 2020. *Buyers' sourcing strategies and suppliers' markups in Bangladeshi garments*. Sustainable Global Supply Chains. https://www.sustainablesupplychains.org/wp-content/uploads/2021/03/CajalGrossiMacchiavelloNoguera.pdf.

Carballo, J., G. I. P. Ottaviano, and C. Volpe Martincus. 2018. The buyer margins of firms' exports. *Journal of International Economics* 112:33-49.

Carvalho, V. M., and A. Tahbaz-Salehi. 2019. Production networks: A primer. *Annual Review of Economics* 11:635-663.

Chaney, T. 2014. The network structure of international trade. *American Economic Review* 104(11):3600-3634.

Christen, P. 2012. *Data matching*. New York: Springer Publishing.

Clausing, K. 2001. Trade creation and trade diversion in the Canada–United States free trade agreement. *Canadian Journal of Economics* 34(3):677-696.

Clausing, K. 2016. The effect of profit shifting on the corporate tax base in the United States and beyond. *National Tax Journal* 69(4):905-934.

Coase, R. H. 1937. The nature of the firm. *Economica* 4:386-405. https://doi.org/10.1111/j.1468-0335.1937.tb00002.x.

Collard-Wexler, A., and J. De Loecker. 2020. *Production function estimation and capital measurement error*. Working paper. https://www.dropbox.com/s/iurfot319loo280/Capital_Measurement_Error_Production_Functions.pdf?dl=0.

Corrado, C., C. Hulten, and D. Sichel. 2009. Intangible capital and US economic growth. *Review of Income and Wealth* 55(3):661-685.

Corrado, C., J. Haskel, and C. Jona-Lasinio. 2017. Knowledge spillovers, ICT and productivity growth. *Oxford Bulletin of Economics and Statistics* 79(4):592-618.

Carol C., C. Hulten, and D. Sichel. 2005. Measuring capital and technology: An expanded framework. In *Measuring capital in the new economy*, edited by C. Corrado, J. Haltiwanger and D. Sichel. Cambridge, MA: National Bureau of Economic Research. https://www.nber.org/system/files/chapters/c0202/c0202.pdf.

Costinot, A., and A. Rodríguez-Clare. 2014. Trade theory with numbers: Quantifying the consequences of globalization. In *Handbook of international economics*, vol. 4, edited by G. Gopinath, E. Helpman, and K. Roggoff. Amsterdam: Elsevier. Pp. 197-261.

Coyle, D., and Nguyen, D. 2020, June 16. No plant, no problem? Factoryless manufacturing, economic measurement and national manufacturing policies. *Review of International Political Economy*. https://doi.org/10.1080/09692290.2020.1778502.

Davies, R., J. Martin, M. Parenti, and F. Toubal. 2018. Knocking on tax haven's door: Multinational firms and transfer pricing. *Review of Economics and Statistics* 100(1):120-134.

De Loecker, J., J. Eeckhout, and G. Unger. 2020. The rise of market power and the macroeconomic implications. *Quarterly Journal of Economics* 135(2):561-644.

Dedrick, J., K. Kraemer, and G. Linden. 2010. Who profits from innovation in global value chains?: A study of the iPod and notebook PCs. *Industrial and Corporate Change* 19(1):81-116.

Dekle, R., J. Eaton, and S. Kortum. 2008. Global rebalancing with gravity: Measuring the burden of adjustment. *IMF Staff Papers* 55(3):511-540.

Demir, B., A. Fieler, D. Xu, and K. Kaili Yang. 2021. *O-ring production networks*. Working Paper w28433. Cambridge, MA: National Bureau of Economic Research.

Derrick, A. & C.P. Steiner. (In progress). *Intellectual property networks within U.S. multinational enterprises*. Working Paper. Suitland, MD: U.S. Bureau of Economic Analysis.

Dharmapala, D. 2014. What do we know about base erosion and profit shifting? A review of the empirical literature. *Fiscal Studies* 35(4):421-448.

Dharmapala, D., and N. Riedel. 2013. Earnings shocks and tax-motivated income-shifting: Evidence from European multinationals. *Journal of Public Economics* 97:95-107.

Dhyne, E., A. Ken Kikkawa, M. Mogstad, and F. Tintelnot. 2021. Trade and domestic production networks. *Review of Economic Studies* 88(2):643-668. https://doi.org/10.1093/restud/rdaa062.

Diewert, W. E. 1976. Exact and superlative index numbers. *Journal of Econometrics* 4(2):115-145.

Dischinger, M., and N. Riedel. 2011. Corporate taxes and the location of intangible assets within multinational firms. *Journal of Public Economics* 95(7-8):691-707.

Dragusanu, R. 2014. *Firm-to-firm matching along the global supply chain*. Dissertation. Cambridge, MA: Harvard University.

Du, L., J. Mao, and J. Shi. 2009. Assessing the impact of regulatory reforms on China's electricity generation industry. *Energy Policy* 37(2):712-720.

Dyer, J. 1996. Does governance matter? Keiretsu alliances and asset specificity as sources of Japanese competitive advantage. *Organization Science* 7(6):649-666.

Eaton, J., and S. Kortum. 2001. Trade in capital goods. *European Economic Review* 45(7):1195-1235.

Eaton, J, M. Eslava, M. Kugler, and J. Tybout. 2007. The margins of entry into export markets: Evidence from Colombia. In *Globalization and the organization of firms and markets*. Conference Proceedings. October 2–November 2, 2007. Munich, Germany.

Eaton, J., S. Kortum, and F. Kramarz. 2011. An anatomy of international trade: Evidence from French firms. *Econometrica* 79(5):1453-1498.

Eaton, J., M. Eslava, D. Jinkins, C. J. Krizan, and J. Tybout. 2014. *A search and learning model of export dynamics.* Working Paper. State College, PA: The Pennsylvania State University.

Eaton, J., S. Kortum, B. Neiman, and J. Romalis. 2016a. Trade and the global recession. *American Economic Review* 106(11):3401-3438.

Eaton, J., D. Jinkins, J. Tybout, and D. Xu. 2016b. Two-sided search in international markets. In *2016 Annual Meeting of the Society for Economic Dynamics.* Conference Proceedings. June 30–July 2, 2016. Toulouse, France.

Eaton, J., M. Eslava, D. Jinkins, C. J. Krizan, and J. Tybout. 2021. *A search and learning model of export dynamics.* Working Paper w29100. Cambridge, MA: National Bureau of Economic Research.

European Commission, International Monetary Fund, Organisation for Economic Co-operation and Development, United Nations, and World Bank. 2009. *System of national accounts 2008.* New York. https://unstats.un.org/unsd/nationalaccount/docs/SNA2008.pdf.

Eurostat. 2017. Globalisation patterns in DU trade and investment. https://ec.europa.eu/eurostat/documents/3217494/8533590/KS-06-17-380-EN-N.pdf/8b3e000a-6d53-4089-aea3-4e33bdc0055c.

Fally, T., and R. Hillberry. 2018. A co-Asian model of international production chains. *Journal of International Economics* 114:299-315.

Feenstra, R. C., and G. H. Hanson. 1999. The impact of outsourcing and high-technology capital on wages: Estimates for the United States, 1979–1990. *Quarterly Journal of Economics* 114(3):907-940.

Ferrantino, M. J., X. Liu, and Z. Wang. 2012. Evasion behaviors of exporters and importers: Evidence from the US–China trade data discrepancy. *Journal of International Economics* 86(1):141-157.

Fetzer, J. J., T. Highfill, K. W. Hossiso, T. F. Howells, III, E. H. Strassner, and J. A. Young. 2021. "Accounting for Firm Heterogeneity within U.S. Industries: Extended Supply-Use Tables and Trade in Value Added using Enterprise and Establishment Level Data." Working Paper 25249. National Bureau of Economic Research. Revised March. https://www.nber.org/system/files/working_papers/w25249/w25249.pdf.

Fisman, R., and S-J. Wei. 2004. Tax rates and tax evasion: Evidence from "missing imports" in China. *Journal of Political Economy* 112(2):471-496.

Fort, T. and S. Klimek. 2018. "The Effects of Industry Classification Changes on US Employment Composition." Working Paper 18-28, Center for Economic Studies.

Fuchs, C. 2014. *Digital labour and Karl Marx.* New York: Routledge.

Furman, J., and P. Orszag. 2018. A firm-level perspective on the role of rents in the rise in inequality. In *Toward a just society.* New York: Columbia University Press. Pp. 19-47.

Ganapati, W. F. Wong, and O. Ziv. 2021. *Entrepôt: Hubs, scale, and trade costs*. Working Paper no. 29015. Cambridge, MA: National Bureau of Economic Research.

Gollop, F. M., B. M. Fraumeni, and D. W. Jorgenson. 1987. *Productivity and US economic growth*. Cambridge, MA: Harvard University Press.

Greenstone, M., R. Hornbeck, and E. Moretti. 2010. Identifying agglomeration spillovers: Evidence from winners and losers of large plant openings. *Journal of Political Economy* 118(3):536-598.

Griffith, D., G. Yalcinkaya, and G. Rubera. 2014. Country-level performance of new experience products in a global rollout: The moderating effects of economic wealth and national culture. *Journal of International Marketing* 22(4):1-20.

Grossman, S. J. and O. D. Hart. 1986. The costs and benefits of ownership: A theory of vertical and lateral integration. *Journal of Political Economy* 94(4)691-719.

Gutiérrez, G., and T. Philippon. 2017. *Declining competition and investment in the US*. Working Paper no. w23583. Cambridge, MA: National Bureau of Economic Research.

Gutiérrez, G., and T. Philippon. 2019. Fading Stars. AEA Papers and Proceedings. 109(May):312-316. DOI: 10.1257/pandp.20191065.

Handley, K., F. Kamal, and R. Monarch. 2020. *Rising import tariffs, falling export growth: When modern supply chains meet old-style protectionism*. Working Paper no. w26611. Cambridge, MA: National Bureau of Economic Research.

Hanushek, E., and L. Woessmann. 2012. Do better schools lead to more growth? Cognitive skills, economic outcomes, and causation. *Journal of Economic Growth* 17(4):267-321.

Hart, O., and J. Moore. 1990. Property rights and the nature of the firm. *Journal of Political Economy* (98)6:1119-1158.

Haskel, J., and S. Westlake. 2018a, January 23. Productivity and secular stagnation in the intangible economy. *Vox*. https://voxeu.org/article/productivity-and-secular-stagnation-intangible-economy.

Haskel, J., and S. Westlake. 2018b. *Capitalism without capital: The rise of the intangible economy*. Princeton, NJ: Princeton University Press. https://doi.org/ 10.2307/j.ctvc77hhj.

Helper, S., T. Krueger, and H. Wial. 2012. *Locating American manufacturing: Trends in the geography of production*. Washington, DC: Brookings.

Helpman, E., M. Melitz, and S. Yeaple. 2004. Export versus FDI with heterogeneous firms. *American Economic Review* 94(1):300-316.

Hines Jr., J. R. 2010. Tax havens. *Journal of Economic Perspectives*, 24(4):103-126.

Hines, J. R., and E. M. Rice. 1994. Fiscal paradise: Foreign tax havens and American business. *Quarterly Journal of Economics* 109(1):149-182.

Houseman, S., and M. Mandel, eds. 2015. *Measuring globalization: Better trade statistics for better policy.* Kalamazoo, MI: W.E. Upjohn Institute for Employment Research. https://doi.org/10.17848/9780880994903.

Houseman, S., and K. Ryder Jr., eds. 2010. *Measurement issues arising from growth of globalization.* Kalamazoo, MI: W.E. Upjohn Institute for Employment Research and National Academy of Public Administration. https://www.upjohn.org/measurement/conference_papers.pdf.

Hulten, C. R. 1973. Divisia index numbers. *Econometrica* 41(6):1017-1025.

Hulten, C. R. 1978. Growth accounting with intermediate inputs. *Review of Economic Studies* 45(3):511-518.

Hulten, C. R. 1979. On the "importance" of productivity change. *American economic Review* 69(1):126-136.

Hummels, D., J. Ishii, and K-M. Yi. 2001. The nature and growth of vertical specialization in world trade. *Journal of International Economics* 54(1):75-96.

Hummels, D., J. Munch, L. Skipper, and C. Xiang. 2012. Offshoring, transition and training: Evidence from Danish matched worker-firm data. *American Economic Review P&P* 102(3):424-428.

Hummels, D., R. Jørgensen, J. Munch, and C. Xiang. 2014. The wage effects of offshoring: Evidence from Danish matched worker-firm data. *American Economic Review* 104(6):1597-1629.

Hummels, D., J. R. Munch, and C. Xiang. 2018. Offshoring and labor markets. *Journal of Economic Literature* 56(3):981-1028.

Huneeus, F. 2018. *Production network dynamics and the propagation of shocks.* Thesis. Princeton, NJ: Princeton University.

Iacovone, L., B. Javorcik, W. Keller, and J. Tybout. 2015. Supplier responses to Walmart's invasion in Mexico. *Journal of International Economics* 95(1):1-15.

Jona-Lasinio, C., and S. Manzocchi. 2012. *Intangible assets and productivity growth differentials across EU economies: The role of ICT and R&D.* Working Paper 102. Rome: Luiss Lab.

IMF (International Monetary Fund). 2009. *Balance of payments and international investment position manual (BPM6)*, 6th ed. Washington, DC. https://www.imf.org/external/pubs/ft/bop/2007/pdf/BPM6.pdf.

Iyoha, E. 2021. *Essays on the role of networks in firm productivity and international trade.* Dissertation. Nashville, TN: Vanderbilt University.

Javorcik, B. S. 2004. Does foreign direct investment increase the productivity of domestic firms?: In search of spillovers through backward linkages. *American Economic Review* 94(3):605-627.

Javorcik, B. S., and G. Narciso. 2017. WTO accession and tariff evasion. *Journal of Development Economics* 125:59-71.

Jenniges, D., R. Mataloni, S. Stutzman, and Y. Xin. 2019. Strategic movement of intellectual property within US multinational enterprises. In *The challenges of globalization in the measurement of national accounts,* edited by N. Ahmad, B. Moulton, J. D. Richardson, and P. van de Ven. Chicago, IL: University of Chicago Press.

Johnson, R. C. 2018. Measuring global value chains. *Annual Review of Economics* 10:207-236.

Jorgenson, D. W. 1963. Capital theory and investment behavior. *American Economic Review* 53(2):247-259.

Jorgenson, D. W., and Z. Griliches. 1967. The explanation of productivity change. *Review of Economic Studies* 34(3):249-283.

Kamal, F., and R. Monarch. 2018. Identifying foreign suppliers in US import data. *Review of International Economics* 26(1):117-139.

Kamal, F., and W. Ouyang. 2020. *Identifying US merchandise traders: Integrating customs transactions with business administrative data.* Working Paper CES-20-28. Washington, DC: Center for Economic Studies, U.S. Census Bureau.

Kamal, F., and A. Sundaram. 2014. *Buyer-seller relationships in international trade: Do your neighbors matter?* Discussion Paper. Washington, DC: Center for Economic Studies, U.S. Census Bureau.

Kellenberg, D., and A. Levinson. 2019. Misreporting trade: Tariff evasion, corruption, and auditing standards. *Review of International Economics* 27(1):106-129.

Keller, W. 2021. *Knowledge spillovers, trade, and foreign direct investment.* Working Paper w28739. Cambridge, MA: National Bureau of Economic Research.

Keller, W., and S. Yeaple. 2009. Multinational enterprises, international trade, and productivity growth: Firm-level evidence from the United States. *Review of Economics and Statistics* 91(4):821-831.

Keller, W., and S. Yeaple. 2013. The gravity of knowledge. *American Economic Review* 103(4):1414-1444.

Kikuchi, T., K. Nishimura, and J. Stachurski. 2018. Span of control, transaction costs, and the structure of production chains. *Theoretical Economics* 13(2):729-760.

Koh, D., R. Santaeulàlia-Llopis, and Y. Zheng. 2020. Labor share decline and intellectual property products capital. *Econometrica* 88(6):2609-2628.

Kohler, W., and M. Smolka. 2014. Global sourcing and firm selection. *Economics Letters* 124(3):411-415.

Koopman, R., Z. Wang, and S-J. Wei. 2012. Estimating domestic content in exports when processing trade is pervasive. *Journal of Development Economics* 99(1):178-189.

Koopman, R., Z. Wang, and S-J. Wei. 2014. Tracing value-added and double counting in gross exports. *American Economic Review* 104(2):459-494.

Kornfeld, R. 2013. Initial results of the 2013 Comprehensive Revision of the National Income and Product Accounts. In *Survey of current business.* Suitland, MD: Bureau of Economic Analysis. Pp. 6-17. https://apps.bea.gov/scb/pdf/2013/08%20August/0813_niparevision%20text.pdf.

Lev, B., and F. Gu 2016. *The end of accounting and the path forward for investors and managers.* Hoboken, NJ: John Wiley & Sons.

Lim, S. 2018. Determinants of the performance of investment promotion agencies: Evidence from a mix of emerging economies. *Emerging Markets Finance and Trade* 54(8):1907-1923.

Liu, E. 2019. Industrial policies in production networks. *Quarterly Journal of Economics* 134(4):1883-1948.

Los, B., M. Timmer, and G. de Vries. 2015. Global value chains: "Factory World" is emerging. In *The age of global value chains: Maps and policy issues*, edited by J. Amador and F. di Mauro. Ebook. London: Centre for Economic Policy Research. Pp. 36-47.

Macchiavello, R., and P. Miquel-Florensa. 2018. *Vertical integration and inter-firm relationships: Evidence from the Costa Rica coffee chain.* Working Paper. London: London School of Economics.

Macchiavello, R., and A. Morjaria. 2015. The value of relationships: Evidence from a supply shock to Kenyan rose exports. *American Economic Review* 105(9):2911-2945.

Macchiavello, R., and A. Morjaria. 2021. Competition and relational contracts in the Rwanda coffee chain. *Quarterly Journal of Economics* 136(2):1089-1143.

Martin, J., I. Mejean, and M. Parenti. 2020. *Relationship stickiness and economic uncertainty.* Discussion Paper 15609. London: Centre for Economic Policy Research. https://cepr.org/active/publications/discussion_papers/dp.php?dpno=15609.

Meade, D. S., S. Rzeznik, and D. Robinson-Smith. 2003. Business investment by industry in the US economy for 1997. *Survey of Current Business* 83(11):18-70.

Miroudot, S., and M. Ye. 2020. Multinational production in value-added terms. *Economic Systems Research* 32(3):395-412.

Mishra, P., A. Subramanian, and P. Topalova. 2008. Tariffs, enforcement, and customs evasion: Evidence from India. *Journal of Public Economics* 92(10-11):1907-1925.

Monarch, R. 2019. "It's not you, it's me": Prices, quality, and switching in US-China trade relationships. Working Paper, Federal Reserve Board.

Monarch, R., and T. Schmidt-Eisenlohr. 2017. *Learning and the value of trade relationships.* International Finance Discussion Paper 1218. Washington, DC: Board of Governors of the Federal Reserve System.

Nagengast, A. J., and R. Stehrer. 2016. Accounting for the differences between gross and value added trade balances. *World Economy* 39(9):1276-1306.

NBER (National Bureau of Economic Research). 2006. *International service flows.* Conference on Research in Income and Wealth. April 28–29, 2006. Bethesda, MD.

NBER. 2018. *The challenges of globalization in the measurement of national accounts.* Conference on Research in Income and Wealth. Bethesda, MD. March 9–10, 2018. https://www.nber.org/conferences/criwconference-challenges-globalization-measurement-national-accounts-spring-2018.

Nielsen, P. B. 2018. The puzzle of measuring global value chains—The business statistics perspective. *International Economics* 153(C):69-79.Noguera, G. 2012. *Trade costs and gravity for gross and value added trade.* Job Market Paper. New York: Columbia University.

Novak, S., and S. Stern. 2008. How does outsourcing affect performance dynamics? Evidence from the automobile industry. *Management Science* 54(12).

OECD (Organisation for Economic Co-operation and Development). 2010. *Handbook on deriving capital measures of intellectual property products.* Paris: OECD Publishing. https://www.oecd.org/sdd/na/44312350.pdf.

OECD. 2015. *Frascati manual 2015: Guidelines for collecting and reporting data on research and experimental development, the measurement of scientific, technological and innovation activities.* Paris: OECD Publishing. http://dx.doi.org/10.1787/9789264239012-en.

OECD. 2017. *OECD transfer pricing guidelines for multinational enterprises and tax administrations 2017.* Paris: OECD Publishing. https://www.oecd-ilibrary.org/taxation/oecd-transfer-pricing-guidelines-for-multinational-enterprises-and-tax-administrations-2017_tpg-2017-en.

Okubo, S., C. Robbins, C. Moylan, B. Sliker, L. Schultz, and L. Mataloni. 2006. *R&D satellite account: Preliminary estimates.* Suitland, MD: Bureau of Economic Analysis, National Science Foundation. http://beagov.prod.acquiasites.com/sites/default/files/newsreleases/general/rd/2006/pdf/rdreport06.pdf.

Olsen, T., and Statistics Denmark. 2008. *Emission permits.* Issue paper prepared for the 13th meeting of the London Group of Environmental Accounting. September 30–October 3, 2008. Brussels: Statistics Denmark.

Orsini, N., and C. A. S. dos Santos. 2015. *United States-Brazil report on merchandise trade statistics report 2012-2014.* Washington, DC: U.S. Census Bureau.

Pietrobelli, C., and R. Rabellotti. 2011. Global value chains meet innovation systems: Are there learning opportunities for developing countries? *World Development* 39(7):1261-1269.

Porter, M. E. Technology and competitive advantage. 1985. *Journal of Business Strategy* 5(3):60-78.

Reinsdorff, M., and M. Slaughter, eds. 2009. International trade in services and intangibles in the era of globalization. In *NBER Studies in Income and Wealth*, vol. 69. Chicago, IL: University of Chicago Press. http://www.nber.org/books/rein09-1.

Riedel, N. 2018. Quantifying international tax avoidance: A review of the academic literature. *Review of Economics* 69(2):169-181.

Rodríguez-Clare, A. 2010. Offshoring in a Ricardian world. *American Economic Journal: Macroeconomics* 2(2):227-258.

Sallusti, F. 2019. *Detecting and measuring BEPS by MNEs in Italy: A micro approach*. Working Paper. Joint Meeting of the Working Party on Financial Statistics and the Working Party on National Accounts. November 5–7, 2019. Paris: Organisation for Economic Co-operation and Development.

Smarzynska Javorcik, B. 2004. Does foreign direct investment increase the productivity of domestic firms? In search of spillovers through backward linkages. *American Economic Review* 94(3):605-627.

Stoyanov, A. 2012. Tariff evasion and rules of origin violations under the Canada-US Free Trade Agreement. *Canadian Journal of Economics* 45(3):879-902.

Sturgeon, T. J. 2013. *Global value chains and economic globalization—Towards a new measurement framework*. Report to Eurostat. Cambridge, MA: Massachusetts Institute of Technology.

Sturgeon, T. J., P. Bøegh Nielsen, G. Linden, G. Gereffi, and C. Brown. 2013. Direct measurement of global value chains: Collecting product-and firm-level statistics on value added and business function outsourcing and offshoring. In *Trade in value added*, edited by A. Mattoo, Z. Wang, and S-J. Wang. Washington, DC: International Bank for Reconstruction and Development/The World Bank. Pp. 289-321.

Sugita, Y., T. Furusawa, A. Jakobsson, and Y. Yamamoto. 2019. *Global value chains and aggregate income volatility*. Tokyo: Hitotsubashi University.

Sugita, Y., K. Teshima, and E. Seira. 2014 *Assortative matching of exporters and importers*. Working Paper. https://assets.publishing.service.gov.uk/media/57a089eded915d3cfd0004c8/Assortative-Matching-Exporters-Importers.pdf.

Surana, K., C. Doblinger, L. Diaz Anadon, and N. Hultman. 2020. Effects of technology complexity on the emergence and evolution of wind industry manufacturing locations along global value chains. *Nature Energy* 5(10):811-821.

Sutton, J. 2004. *The auto-component supply chain in China and India—A benchmark study*. STICERD Research Paper EI34. London: London School of Economics and Political Science.

Tomiura, E. 2007. Foreign outsourcing, exporting, and FDI: A productivity comparison at the firm level. *Journal of International Economics* 72(1):113-127.

Tørsløv, T. R., L. S. Wier, and G. Zucman. 2018. *The missing profits of nations*. Working Paper w24701. Cambridge, MA: National Bureau of Economic Research.

UNECE (United Nations Economic Commission for Europe). 2011. *The impact of globalization on national accounts*. New York and Geneva: United Nations. https://unece.org/DAM/stats/publications/Guide_on_Impact_of_globalization_on_national_accounts__web_.pdf.

UNECE. 2015. *Guide to measuring global production.* New York and Geneva: United Nations. https://unece.org/statistics/publications/guide-measuring-global-production-2015.
UN (United Nations), International Monetary Fund, Organisation for Economic Co-operation and Development, Statistical Office of the European Union, United Nations Conference on Trade and Development, World Tourism Organization, World Trade Organization. 2012. *Manual on statistics of international trade in services 2010.* New York: United Nations. https://unstats.un.org/unsd/tradeserv/TFSITS/manual.htm.
Upjohn (W.E. Upjohn Institute for Employment Research). 2013. *Measuring the effects of globalization.* Conference. February 28 and March 1. Washington, DC.
Upjohn (W.E. Upjohn Institute for Employment Research) and NAPA (National Academy of Public Administration). 2009. *Measurement issues arising from growth of globalization.* Conference. November 6–7, 2009. Washington, DC.
U.S. Census Bureau. 1996. *Reconciliation of the 1993 and 1994 merchandise trade statistics of the United States and Australia.* Washington, DC: U.S. Census Bureau.
U.S. Census Bureau. 2012. *The second phase report on the reconciliation of the merchandise trade statistics of the United States and China.* Washington, DC: U.S. Census Bureau. https://www.census.gov/foreign-trade/aip/recon_china_080910.pdf.
vom Lehn, C., and T. Winberry. 2022. The investment network, sectoral comovement, and the changing US business cycle. *Quarterly Journal of Economics* 137(1):387-433. https://doi.org/10.1093/qje/qjab020.
Wang, Z., S. Wei, X. Yu, and K. Zhu. 2017. *Measures of participation in global value chains and global business cycles.* Working Paper w23222. Cambridge, MA: National Bureau of Economic Research.
Williamson, O. E. 1975. *Markets and hierarchies: Analysis and antitrust implications: A study in the economics of internal organization.* Glencoe, IL: Free Press.
Williamson, O. 1979. Transaction-cost economics: The governance of contractual relations. *Journal of Law and Economics* 22(2):233-261.
Wooldridge, J. 2009. On estimating firm-level production functions using proxy variables to control for unobservables. *Economics Letters* 104(3):112-114.

APPENDIXES

Appendix A

Workshop Agenda

DAY 1:
Wednesday – May 5, 2021
Via Zoom
All Times Are USA Eastern Time (EDT)

10:00 AM **Welcome**
Susan Helper, Case Western Reserve University (Cochair*)
Wolfgang Keller, University of Colorado Boulder (Cochair*)

10:15 AM **Paper: Multinational Firms and Global Innovation**
Anna Gumpert (LMU Munich), Kalina Manova (University College London), Cristina Rujan (Max Planck Institute), Monika Schnitzer (LMU Munich)

Moderator: Justin R. Pierce, Federal Reserve Board*
Discussant: *Allison Derrick, U.S. Bureau of Economic Analysis*

11:05 AM **Paper: Tracing Value Added in the Presence of Multinational Firms with an Application to High-Tech Sectors**
Zhi Wang (George Mason University); *Shang-Jin Wei* (Columbia Business School, NBER, CEPR); Xinding Yu (University of International Business and Economics); and Kunfu Zhu (Renmin University)

* Member of the Planning Committee

Moderator: Sally Thompson, U.S. Bureau of Economic Analysis (retired)*
Discussant: *Thomas F. Howells III, U.S. Bureau of Economic Analysis*

11:55 AM Break

12:15 PM Paper: **Trade in Services, Intangible Capital, and the Profit-Shifting Hypothesis**
Nadia Accoto, Stefano Federico, and *Giacomo Oddo* (Bank of Italy)

Moderator: Nadim Ahmad, Organisation of Economic Co-operation and Development*
Discussant: *James Hines, University of Michigan*

1:05 PM Paper: **Talent, Geography, and Offshore R&D**
Jingting Fan (Penn State University)

Moderator: Eduardo Morales, Princeton University*
Discussant: *Gary Lyn, Iowa State University*

1:55 PM Break

2:15 PM Paper: **The Nature and Direction of Innovation in Global Value Chains for Wind-Energy Technologies**
Kavita Surana (University of Maryland, College Park), *Claudia Doblinger* (Technical University of Munich), Deyu Li (University of Cambridge), Nathan Hultman (University of Maryland, College Park), Laura Diaz Anadon (University of Cambridge)

Moderator: Eduardo Morales, Princeton University*
Discussant: *Kelly Sims Gallagher, Tufts University*

3:05 PM Paper: **Economies of Scope and Relational Contracts: Exploring Global Value Chains in the Automotive Industry**
Susan Helper (Case Western Reserve University)* and Abdul Munasib (U.S. Bureau of Economic Analysis)

Moderator: *Wolfgang Keller, University of Colorado Boulder (Cochair*)*
Discussant: *Davin Chor, Dartmouth College**

---

* Member of the Planning Committee

APPENDIX A

3:55 PM   **Day 1 Closing Remarks**
Susan Helper, Case Western Reserve University (Cochair*)
Wolfgang Keller, University of Colorado Boulder (Cochair*)

4:00 PM   **Adjourn Day 1**

DAY 2:
Thursday – May 6, 2021
Via Zoom
All Times Are USA Eastern Time (EDT)

9:30 AM   **Welcome**
Susan Helper, Case Western Reserve University (Cochair*)
Wolfgang Keller, University of Colorado Boulder (Cochair*)

9:35 AM   **Keynote Address:**
Foreign Direct Investments and Superstar Spillovers: Evidence from Firm-to-Firm Transactions
Mary Amiti (Federal Reserve Bank of New York and CEPR), Cedric Duprez (National Bank of Belgium), Jozef Konings (Nazarbayev University and KU Leuven), *John Van Reenen* (London School of Economics)

10:05 AM   **Paper: Creation and Diffusion of Knowledge in the Global Firm**
Çağatay Bircan (EBRD), Beata Javorcik (EBRD, University of Oxford, and CEPR), and Stefan Pauly (Sciences Po)

Moderator: Andreas Moxnes, University of Oslo*
Discussant: *Heiwai Tang, Hong Kong University*

10:55 AM   **Paper: Firm Selection and Organizational Choice: Complex Patterns of Global Sourcing**
Valérie Smeets and *Frédéric Warzynski* (Aarhus University)

Moderator: Davin Chor, Dartmouth College*
Discussant: *Catherine Thomas, London School of Economics*

11:45 AM   **Break**

---

* Member of the Planning Committee

12:05 PM **Paper: Are Customs Records Consistent Across Countries?**
C. J. Krizan (U.S. Department of Labor), *James Tybout* (Penn State University and NBER), Zi Wang (Shanghai University of Finance and Economics), and Yingyan Zhao (George Washington University)

Moderator: Andreas Moxnes, University of Oslo*
Discussant: *Jeronimo Carballo, University of Colorado Boulder*

12:55 PM **Paper: Capital Flows in Global Value Chains**
Xiang Ding (Georgetown University)

Moderator: Nadim Ahmad, Organisation of Economic Co-operation and Development*
Discussant: *Brent Moulton, International Monetary Fund*

1:45 PM Break

2:05 PM **Paper: Colocation of Production and Innovation: Evidence from the United States**
Teresa C. Fort (Tuck School at Dartmouth College, CEPR, and NBER), Wolfgang Keller (University of Colorado Boulder, CEPR, and NBER),* Peter K. Schott (Yale School of Management, CEPR, and NBER), Stephen Yeaple (Penn State University and NBER), and Nikolas Zolas (Center for Economic Studies, U.S. Census Bureau)

Moderator: Justin R. Pierce, Federal Reserve Board*
Discussant: *Nicholas Bloom, Stanford University*

2:55 PM **Day 2 Closing Remarks**
Susan Helper, Case Western Reserve University (Cochair)*
Wolfgang Keller, University of Colorado Boulder (Cochair)*

3:00 PM **Adjourn Day 2**

* Member of the Planning Committee

DAY 3:
Friday – May 7, 2021
Via Zoom
All Times are USA Eastern Time (EDT)

10:00 AM **Welcome**
Susan Helper, Case Western Reserve University (Co-chair*)
Wolfgang Keller, University of Colorado (Co-chair*)

10:05 AM **Panel: Global Value Chain Measurement Methodology: Challenges and Prospects**
Moderator: Sally Thompson, U.S. Bureau of Economic Analysis (retired)*

Measuring Global Value Chains
Timothy J. Sturgeon, Massachusetts Institute of Technology

Quality Challenges in Modernizing Official Business Statistics
Oscar Lemmers, Statistics Netherlands

Measuring R&D and Competitiveness of U.S. Value-Added Exports: Prospects and Challenges
Jon D. Samuels, U.S. Bureau of Economic Analysis

Discussant: Francisco Moris, National Science Foundation

11:15 AM Break

11:30 AM **Lessons from the Workshop: A Panel Discussion**
Moderator: Susan Helper, Case Western Reserve University (Cochair*)

Rebecca Riley, ESCoE and King's College London
William Powers, U.S. International Trade Commission
Maria Borga, International Monetary Fund
Nadim Ahmad, Organisation for Economic Co-operation and
  Development*

12:30 PM **Closing Remarks**
Wolfgang Keller, University of Colorado Boulder (Cochair*)

12:45 PM **Adjourn**

* Member of the Planning Committee

Appendix B

Biographies of Speakers and Planning Committee Members (as of May 2021)

NADIM AHMAD
(Member of the Planning Committee)

Nadim Ahmad is deputy director at the Organisation for Economic Co-operation and Development (OECD) Centre for Entrepreneurship, SMEs, Regions and Cities (CFE), helping to drive momentum in the development of integrated policies that look holistically at people, places, and firms—promoting better policies for better lives, and resilient and sustainable economic growth. In his role, Mr. Ahmad provides intellectual leadership and quality control to ensure that OECD is at the forefront of policy thinking in the domains covered by CFE. Before joining CFE in 2020, he worked in OECD's Statistics and Data Directorate, where he led international efforts to better account for globalization; entrepreneurship and business performance; and, in particular, the role of multinational enterprises and small- and medium-sized enterprises in global value chains. Prior to that he worked with OECD's Directorate for Science Technology and Innovation, where he developed OECD's first estimates of carbon dioxide emissions embodied in international trade. Prior to joining the OECD in 2000, Mr. Ahmad worked in the United Kingdom's Office for National Statistics (1996–2000) and Ministry of Finance (1993–1996).

ÇAĞATAY BIRCAN

Çağatay Bircan is a senior research economist at the Office of the Chief Economist at the European Bank for Reconstruction and Development (EBRD) in London. He conducts academic research and various types of policy work at the EBRD. He researches topics at the intersection of international economics, finance, and development. Currently, he is exploring topics on innovation, knowledge work, and productivity in multinational firms and private equity. Dr. Bircan earned a B.A. in economics, mathematics, and German from Williams College, and an M.A. and a Ph.D. in economics from the University of Michigan.

NICHOLAS (NICK) BLOOM

Nicholas (Nick) Bloom is the William Eberle professor of economics at Stanford University, a senior fellow of the Stanford Institute for Economic Policy and Research, and codirector of the Productivity, Innovation and Entrepreneurship program at the National Bureau of Economic Research. His research focuses on management practices and uncertainty. Previously, Dr. Bloom worked at the U.K. Treasury and McKinsey & Company. He is a fellow of the American Academy of Arts and Sciences, and the recipient of the Sloan Fellowship, the Bernacer Prize, the Frisch Medal, and a National Science Foundation Career Award. Dr. Bloom earned a B.A. from Cambridge University, an M.Phil. from Oxford University, and a Ph.D. from University College London.

MARIA BORGA

Maria Borga is deputy division chief in the Balance of Payments Division at the International Monetary Fund (IMF), where she is playing a lead role in the update of the international statistical standards, including the *Balance of Payments and International Investment Position Manual*. Prior to joining the IMF, she served as senior statistician at the Organisation for Economic Co-operation and Development (OECD), where she oversaw the implementation of OECD's *Benchmark Definition of Foreign Direct Investment, 4th edition* (*BD4*). Before joining OECD, Dr. Borga served as assistant division chief for research and analysis in the Balance of Payments Division of the U.S. Bureau of Economic Analysis (BEA), where she led research into improving BEA's measures of banking, wholesale and retail trade, and insurance services to better capture trade in these services. She received a B.A. in economics and a B.S. in French from The Pennsylvania State University and a Ph.D. in economics from Boston University.

JERONIMO CARBALLO

Jeronimo Carballo is an assistant professor in the Department of Economics at the University of Colorado Boulder, where he has served since 2015. His primary research interest is international trade with a special emphasis on firms' decisions under uncertainty, exploring trade cost using microdata, and the intersection between trade and labor markets. His research has been published in several journals such as the *Journal of International Economics* and *Journal of Development Economics*. Dr. Carballo earned a B.S. in economics from the National University of Cordonba, Argentina; an M.A. in economics from the National University of La Plata, Argentina; and a Ph.D. in economics from the University of Maryland.

DAVIN CHOR
(Member of the Planning Committee)

Davin Chor is an associate professor and globalization chair at the Tuck School of Business, Dartmouth College. His research interests are in international trade and political economy. As part of Dartmouth's academic cluster on globalization, Dr. Chor studies the far-reaching repercussions of globalization on world markets, governments, trade, and society. He is a research associate of the National Bureau of Economic Research (NBER) and presently serves as an associate editor at the *Journal of International Economics* and at the *Review of International Economics*. Dr. Chor completed his A.B. in economics summa cum laude from Harvard University in 2000. He also holds an A.M. in statistics and a Ph.D. in economics (2007) from Harvard University.

ALLISON DERRICK

Allison Derrick is a research economist at the U.S. Bureau of Economic Analysis (BEA). She studies multinational enterprises and their role in the U.S. and global economies. Dr. Derrick's main areas of research are intellectual property, research and development, and emerging technologies. She also studies the economics of blockchain technology and crypto assets, including the conceptual and practical challenges of measuring them in economic statistics. Dr. Derrick received her master's and Ph.D. in agricultural and applied economics from the University of Wisconsin–Madison.

XIANG DING

Xiang Ding is an assistant professor in the School of Foreign Service at Georgetown University. His research lies at the intersection of international trade, technology, and policy. One strand of his research leverages microdata from the U.S. Census Bureau to uncover and estimate the aggregate implications of globalization, such as economies of scope from joint production, and structural transformation within U.S. firms. Dr. Ding's research has been recognized by the World Trade Organization Essay Award for Young Economists. He received a Ph.D. in business economics from Harvard University and an A.B. in economics from Princeton University.

CLAUDIA DOBLINGER

Claudia Doblinger is an assistant professor of innovation and technology management at the Technical University of Munich (TUM), where she primarily works on clean-energy innovation and entrepreneurship. Her main focus is on understanding how political incentives affect the innovation and entrepreneurial activities of firms, especially in the context of clean-energy and transportation technologies. Dr. Doblinger's research has been published in peer-reviewed journals such as *Nature Energy*, *Research Policy*, and the *Journal of Product Innovation Management*. She has previously worked at the Harvard Kennedy School, the University of Regensburg (Germany), and a German energy company. Dr. Doblinger holds a Ph.D. in innovation and technology management.

JINGTING FAN

Jingting Fan is an assistant professor in the Department of Economics at The Pennsylvania State University. His research focuses on international trade and urban economics. One strand of his work is studying the impacts of multinational corporations (MNCs) on the income of countries, including investigating MNCs' decisions on financing and carrying out R&D overseas. He also studies within-country spatial friction and its interaction with policies and economic shocks. Dr. Fan earned a B.S. in pure and applied mathematics and a B.A. in economics and finance, both from Tsinghua University, China, and a Ph.D. in economics from the University of Maryland at College Park.

TERESA FORT

Teresa Fort is an associate professor of business administration at the Tuck School of Business at Dartmouth College. She conducts research in international trade and industrial organization. Currently, she analyzes how technology affects firm-level offshoring and production fragmentation decisions, and the impact of these decisions on domestic employment and innovation. Dr. Fort is a faculty research fellow at the National Bureau of Economic Research and a research affiliate at the Centre for Economic Policy Research. She holds a B.A. from the University of Virginia and a Ph.D. in economics from the University of Maryland.

KELLY SIMS GALLAGHER
(Member of the Planning Committee)

Kelly Sims Gallagher is academic dean and professor of energy and environmental policy at The Fletcher School at Tufts University. She directs the Climate Policy Lab and the Center for International Environment and Resource Policy at Fletcher. From June 2014 to September 2015 Dr. Sims Gallagher served in the Obama administration as a senior policy advisor in the White House Office of Science and Technology Policy, and as senior China advisor in the Special Envoy for Climate Change office at the U.S. Department of State. Her most recent book is *Titans of the Climate* with Xuan Xiaowei (MIT Press, 2019). Dr. Sims Gallagher earned an A.B. from Occidental College, and an M.A. in law and diplomacy and a Ph.D. from The Fletcher School at Tufts University.

ANNA GUMPERT

Anna Gumpert is an assistant professor at the Department of Economics of Ludwig-Maximilians-Universität München (LMU Munich). Her research focuses on strategic decisions of firms—in particular, decisions on globalization, organization, and innovation. Dr. Gumpert is a research affiliate of the Centre for Economic Policy Research and CESifo, Munich. She received her Ph.D. in economics from LMU Munich and was a visiting fellow at Harvard University and Yale University.

SUSAN HELPER
(Planning Committee Cochair)

Susan Helper is Frank Tracy Carlton professor of economics at Weatherhead School of Management, Case Western Reserve University. She served as chief economist of the U.S. Department of Commerce from 2013 to 2015, and as senior economist at the White House Council of Economic Advisors from 2012 to 2013. Her research focuses on how global supply chains affect regional development and innovation. Dr. Helper received her B.A. from Oberlin College and her Ph.D. from Harvard University.

JAMES R. HINES JR.

James Hines is Richard A. Musgrave collegiate professor of economics, L. Hart Wright collegiate professor of law, and research director of the Office of Tax Policy Research, all at the University of Michigan. He researches various aspects of taxation. Dr. Hines has also taught at Princeton University and Harvard University, and has held visiting appointments at Columbia University; the London School of Economics; the University of California, Berkeley; and Harvard Law School. In 2017 he received the National Tax Association's Daniel M. Holland Medal for lifetime achievement in the study of public finance. Dr. Hines is a research associate of the National Bureau of Economic Research, research director of the International Tax Policy Forum, former coeditor of the American Economic Association's *Journal of Economic Perspectives*, and a former economist in the U.S. Department of Commerce. He holds a B.A. and M.A. from Yale University and a Ph.D. from Harvard, all in economics.

THOMAS F. HOWELLS

Thomas F. Howells is chief of the Industry Economics Division at the U.S. Bureau of Economic Analysis (BEA), where he oversees preparation of BEA's industry economic accounts, including supply-use tables, statistics on gross domestic product by industry, and other industry products. These accounts provide a comprehensive picture of the performance of industries within the U.S. economy, including how these industries interact and contribute to economic growth. In addition to work on these core industry statistics, Mr. Howells has led several other cross-cutting efforts, including international collaboration on statistics for trade in value added for the Asia-Pacific Economic Cooperation and North America regions, work on the integrated BEA/Bureau of Labor Statistics industry-level production account, and development of BEA's outdoor-recreation satellite account. He has received various awards for his contributions to BEA's accounts, including two U.S. Department of Commerce Silver Medals. Mr. Howells holds a B.A. in economics from Brigham Young University and an M.A. in economics from George Mason University.

WOLFGANG KELLER
(Planning Committee Cochair)

Wolfgang Keller is professor of economics and director of the McGuire Center at the University of Colorado Boulder. His interests are in international trade and investment and economic development, especially the international diffusion of technological knowledge. In the past, Dr. Keller held visiting positions at Stanford, Princeton, and Brown Universities, and he has been a resident scholar at the International Monetary Fund's Research Department, as well as an advisor to both the World Bank and the World Trade Organization. Since 2005, his research has been supported continuously by grants from the National Science Foundation. Dr. Keller is member of both the National Bureau of Economic Research and the Centre for Economic Policy Research, and his research has been published in such journals as the *American Economic Review* and the *Journal of Political Economy*. He received his Diploma degree from the University of Freiburg and his Ph.D. from Yale University.

OSCAR LEMMERS

Oscar Lemmers works as a senior researcher at Statistics Netherlands. For the last 14 years, his work has included measuring globalization by developing new methods and data, and publishing insights about causes and effects related to this phenomenon. His publications have included short web articles, reports for policy makers, and publications in academic journals. Dr. Lemmers also contributes to measuring globalization by serving as chair of the Organisation for Economic Co-operation and Development expert group on extended supply-use tables. This work combines the advantages of national accounts statistics (featuring internal consistency and relevance to macroeconomic totals, such as gross domestic product) with those of detailed business statistics to obtain new data and insights about the role of small- and medium-sized or multinational enterprises in global value chains. Dr. Lemmers holds a Ph.D. in mathematics.

GARY LYN

Gary Lyn is assistant professor of economics at Iowa State University. His current research is in international trade, economic geography, international migration, and the intersection of international trade and environmental economics. Dr. Lyn's principal interests include understanding the effects of external economies of scale in production on the gains from trade and intercountry trade patterns; characterizing the equilibrium properties of within-country trade patterns and factor reallocation in the presence of these external economies; and estimating and quantifying the welfare effect of global warming and climate change, while fully accounting for intra- and intercountry spatial trade and migration linkages. He holds a B.Sc. in economics and an M.Sc. in economics and business from The University of the West Indies, and a Ph.D. in economics from The Pennsylvania State University.

EDUARDO MORALES
(Member of the Planning Committee)

Eduardo Morales is professor of economics at Princeton University, a former assistant professor of economics at Columbia University, and a former fellow in international economics at Princeton University. He researches such topics as innovation in the global firm, the segregation of urban consumption patterns, changes in export behavior, and the effects of international trade and technology on occupational inequality. Dr. Morales has received honors in teaching and scholarship from Harvard University and the Spanish Ministry of Education; he is associate editor of two economic journals. He received his B.A. in economics and law from the Universidad Carlos III de Madrid, and his master's degree and Ph.D. in economics from Harvard University.

FRANCISCO MORIS

Francisco Moris is a senior analyst at the National Science Foundation, National Center for Science and Engineering Statistics. His research interests include methodological and empirical work on globalization measurement, intangibles in national economic accounts, and metadata quality. His work currently supports the incorporation of international property products in input-output, supply-use, and trade-in-value-added statistics in a joint project with the Bureau of Economic Analysis. Dr. Moris has worked on multiple interagency projects linking foreign direct investments, multinational enterprises, research and development (R&D), and services trade microdata. He represented the United States in a United Nations Statistical Commission Working Group on SDMX (Statistical Data and Metadata Exchange) of the Inter-agency and Expert Group on Sustainable Development Goal Indicators. Dr. Moris has published in the *Journal of Industry, Competition and Trade* and in two National Bureau of Economic Research books in the *Studies in Income and Wealth* series. He is a member of the Academy of International Business and of the editorial review board for the *Journal of International Business Policy*. Dr. Moris holds a Ph.D. from George Washington University.

BRENT MOULTON

Brent Moulton is a senior economist at the International Monetary Fund. Previously he spent 19 years as associate director for national economic accounts at the Bureau of Economic Analysis, where he contributed to the 2008 update of the System of National Accounts; he also served for 13 years at the Bureau of Labor Statistics (BLS), most recently as chief of price and index-number research, where he was responsible for the research program to maintain accuracy of the BLS price index and consumer expenditure statistics. Dr. Moulton holds a B.A. and an M.S. in economics from Brigham Young University, and a Ph.D. in economics from the University of Chicago.

ANDREAS MOXNES
(Member of the Planning Committee)

Andreas Moxnes is professor of economics at the University of Oslo. Among his main research interests are the interaction between globalization and the behavior of firms, reallocations in product markets and productivity growth, and the microeconometrics of trade and multinational production. Prior to joining the University of Oslo, Dr. Moxnes was an assistant professor at Dartmouth College, an economist at the Norwegian Ministry of Finance, and a visiting professor at Princeton University for 2019–2020. He has published articles on networks and trade, international sourcing, and the geography of production networks. Dr. Moxnes is editor of the *Scandinavian Journal of Economics* and associate editor of *The Journal of International Economics*. He is on the editorial board of the *Review of Economic Studies*. Dr. Moxnes received his Ph.D. in economics from the University of Oslo.

GIACOMO ODDO

Giacomo Oddo is an economist in the Balance of Payments Division of the Bank of Italy, within the Directorate General for Economics, Statistics, and Research. He works on topics related to multinational firms, international trade, cross-border investment, and financial flows. Mr. Oddo was a visiting scholar at the Economics Department of Columbia University for 2015–2016. He holds master's degrees in economics from the University of Pisa and from the Sant'Anna School of Advanced Studies.

JUSTIN R. PIERCE
(Member of the Planning Committee)

Justin R. Pierce is principal economist with the Federal Reserve Board, a position he has held since 2016. Previously, he was senior economist and economist with the Federal Reserve in the Industrial Output section of the Division of Research and Statistics. Dr. Pierce's expertise is in international trade and industrial organization, and he has contributed research on trade liberalization, the effects of tariffs on U.S. manufacturing, and the decline in manufacturing employment. He received his Ph.D. in economics from Georgetown University.

WILLIAM M. POWERS

William M. Powers is chief economist and director of the Office of Economics at the U.S. International Trade Commission. In this role he serves as the Commission's chief economic adviser and directs economic analysis in the Commission's fact-finding investigations for the U.S. Congress and President. He also leads antidumping and countervailing duty investigations. Dr. Powers is a leader in international efforts to better measure the economic impact of globalization, and was the U.S. chair for the recently completed Asia-Pacific Economic Cooperation Trade-in-Value-Added initiative. He has published on empirical trade topics including trade agreements, global value chains, rules of origin, and trade finance. During the global trade downturn of 2008–2009, Dr. Powers served in a 1-year position as a senior international economist at the President's Council of Economic Advisers (CEA) in both the Bush and Obama administrations. He holds a B.S. in electrical engineering from the University of Virginia and a Ph.D. in economics from the University of Michigan.

REBECCA RILEY

Rebecca Riley is director of the U.K. Economic Statistics Centre of Excellence (ESCoE) and professor of practice in economics at King's Business School, King's College London. She has written extensively on U.K. productivity performance and labor market policy. Ms. Riley is a fellow of the National Institute of Economic and Social Research, where she led the productivity research group and the U.K. forecast team. She has been an external advisor to several government departments and public bodies, including the Department for Work and Pensions and the Office for National Statistics. Ms. Riley is a member of council of the International Association for Research in Income and Wealth, and a member of the Economic and Social Research Council's Centre for Macroeconomics, as well as the Conference on Research in Income and Wealth.

JON SAMUELS

Jon Samuels is a senior research economist at the Bureau of Economic Analysis (BEA) and an associate at the Institute of Quantitative Social Science at Harvard University. His research is focused on measuring productivity and the sources of economic growth, leading to the publication of the BEA/Bureau of Labor Statistics Integrated Industry production account, which presents information on the sources of U.S. economic growth from the bottom-up across industries. Previously, Dr. Samuels worked as a researcher for the Program on Technology and Economic Policy directed by Dale Jorgenson at Harvard University and was a junior economist in economic forecasting with Primark Decision Economics. He received a B.A. in economics from the University of Chicago, and an M.A. and Ph.D. in economics from The Johns Hopkins University.

TIMOTHY J. STURGEON

Timothy J. Sturgeon is a senior researcher at the Industrial Performance Center at the Massachusetts Institute of Technology. His research focuses on the processes of global integration and digital transformation, with an emphasis on offshoring and outsourcing practices in the electronics, automotive, and services industries. Dr. Sturgeon's work explores how evolving technologies and business models are altering linkages between industrialized and developing economies, and development experiences more broadly. His most recent book is *Compressed Development: Time and Timing in Economic and Social Development*, published by Oxford University Press in October 2020. Dr. Sturgeon has been a leader in the interdisciplinary field of global value chains (GVCs) for nearly 20 years, and he has contributed to the development of an assessment of the U.S. data resources available for the study of the international sourcing of services, a research program and classification based on the concept of business functions, and a complementary grouping of trade in services enabled by information and communications technologies. He has also developed a classification of final and intermediate goods in the electronics, motor vehicle, and textile-apparel industries. His work has been used by the World Bank to create a public-use dataset of GVC trade. Dr. Sturgeon has collaborated with the National Science Foundation, Eurostat, Statistics Denmark, and the United Nations Statistics Division, among others.

KAVITA SURANA

Kavita Surana is an assistant research professor at the Center for Global Sustainability at the University of Maryland. Her research is focused on innovation in clean-energy technologies and the interactions between public policy and decision making in private industry to address global and local environmental challenges. Dr. Surana was previously a research associate and postdoctoral fellow in the Science, Technology, and Public Policy Program at the Belfer Center for Science and International Affairs at the Harvard Kennedy School of Government. She has worked with the World Bank, ICF International, and the French Alternative and Atomic Energies Commission (CEA), analyzing and advising on a range of issues related to innovation and energy. Dr. Surana holds a B.Sc. in physics from St. Stephen's College, Delhi University, India; an M.S. in energy storage and conversion materials from Paul Sabatier University, Toulouse, France; and a Ph.D. in materials science and engineering from the Institut Polytechnique de Grenoble (INP-Grenoble) and CEA.

HEIWAI TANG

Heiwai Tang is professor of economics, as well as associate director of the Asia Global Institute, the Institute for China and Global Development, and the Hong Kong Institute of Economics and Business Strategy at the University of Hong Kong (HKU). Prior to joining HKU, he was tenured associate professor of international economics at the School of Advanced International Studies (SAIS) of The Johns Hopkins University. Dr. Tang is also affiliated with the Federal Reserve Bank of Dallas (U.S.), CESIfo (Germany), the Kiel Institute for the World Economy (Germany), and the Globalization and Economic Policy Center (United Kingdom) as a research fellow. He has been a consultant to the World Bank, the International Finance Corporation, the United Nations, and the Asian Development Bank; and he has held visiting positions at the International Monetary Fund, Stanford University, the Massachusetts Institute of Technology (MIT), and Harvard University. Dr. Tang is currently an associate editor of the *Journal of International Economics,* the *Journal of Comparative Economics*, and the *China Economic Review*. He holds a B.S. in mathematics from the University of California, Los Angeles, and a Ph.D. in economics from MIT.

CATHERINE THOMAS

Catherine Thomas is an associate professor of managerial economics and strategy at the London School of Economics. She is director of the International Trade Programme at the Centre for Economic Performance and a research fellow of the Centre for Economic Policy Research. Dr. Thomas was formerly an assistant professor of economics at Columbia Business School. Her research is focused on three aspects of international economic integration: (1) how firms engage in offshoring, (2) how firms make outsourcing decisions, and (3) the performance consequences of firms' organisational structures. Dr. Thomas holds an M.A. in economics from the University of Edinburgh and an M.A. and Ph.D. in business economics from Harvard University.

SALLY THOMPSON
(Member of the Planning Committee)

Dr. Sarahelen "Sally" Thompson is former deputy director of the U.S. Bureau of Economic Analysis (BEA), a position she held from 2016 until her retirement in 2018. As deputy director, she worked to improve and expand BEA's statistical programs. Prior to serving as deputy director, Dr. Thompson was the top executive overseeing international economic statistics at BEA; she led the most significant restructuring of BEA's international statistics since 1976, played a key role in developing statistics that track new direct investment by foreigners in the United States, and oversaw the creation of a data tool to provide easier access to BEA's vast array of international statistics. Prior to her time at BEA, Dr. Thompson was director of the Market and Trade Economics Division at the U.S. Department of Agriculture's Economic Research Service. She has also worked in academia, including as a professor and head of the Department of Agricultural Economics at Purdue University and as a professor of agricultural and consumer economics at the University of Illinois. Dr. Thompson holds a B.A. and M.A. from the University of Minnesota, and a Ph.D. from Stanford University.

JAMES TYBOUT

James Tybout is professor of economics at The Pennsylvania State University and a research associate of the National Bureau of Economic Research. He previously served on the faculty at Georgetown University. Dr. Tybout's recent research has focused on the effects of international trade on the industrial sectors and labor markets of developing countries. He received a bachelor's degree in mathematics and economics from Vanderbilt University, and a master's degree and Ph.D. in economics from the University of Wisconsin–Madison.

JOHN VAN REENEN

John Van Reenen is Ronald Coase School professor at the London School of Economics and digital fellow in the Institute for the Digital Economy at the Massachusetts Institute of Technology (MIT). Until 2020, he was Gordon Billard professor in the MIT Economics Department and Sloan Management School. Dr. Van Reenen has published more than 100 papers on many areas in economics, with a particular focus on firm performance and the causes and consequences of innovation. He was the 2009 winner of the Yrjö Jahnsson Award (the European equivalent of the Clark Medal), the Arrow Prize (2011), the European Investment Bank Prize (2014), and the Harvard Business Review–McKinsey Award (2018). Dr. Van Reenen is a fellow of the British Academy, the Econometric Society, the National Bureau of Economic Research, the Centre for Economic Policy Research, and the Society of Labor Economists. In 2017, he was awarded an Officer of the Order of the British Empire for "services to public policy and economics" by the Queen.

FREDERIC WARZYNSKI

Frederic Warzynski is professor in the Department of Economics and Business Economics at Aarhus University. His research interests include industrial organization, international economics, and organizational economics. He earned a Ph.D. in economics from the Katholieke Universities Leuve.

SHANG-JIN WEI

Shang-Jin Wei is professor of finance and economics at Columbia University's Graduate School of Business and School of International and Public Affairs. From 2014 to 2016, he served as chief economist of the Asian Development Bank and director general of its Economic Research and Regional Cooperation Department. Prior to his appointment at Columbia, Dr. Wei was assistant director and chief of the Trade and Investment Division at the International Monetary Fund. He has also served as associate professor of public policy at Harvard University's Kennedy School of Government. Dr. Wei is a noted scholar on international finance, trade, macroeconomics, and China, with publications in top academic journals, including *American Economic Review, Journal of Political Economy, Quarterly Journal of Economics,* and many others. His research has been reported in popular media, including *Financial Times, Wall Street Journal, The Economist, BusinessWeek,* and more. Dr. Wei earned a B.A. in world economy from Fudan University; an M.A. in economics from The Pennsylvania State University; and an M.S. in business administration (finance) and a Ph.D. in economics from the University of California, Berkeley.

Appendix C
Crosswalk of Workshop Papers to Measurement and Understanding of Global Value Chains

Chapter	Data Sources	Measurement and Valuation Issues	Understanding GVCs and Policy Issues
2. Multinational Firms and Global Innovation	Orbis data (Bundesbank's MiDi in future work); EPO's PATSTAT (German firms only)	Merging these two databases provides inventor location to determine whether production and innovation are colocated; patent data provides information on the type (basic, applied product, or applied process) and quality of innovation.	Better understanding of production and innovation location decisions by MNEs; future work can examine changes in IP rights, tax policy, and FDI policy on MNE production and innovation decisions.
3. Tracing Value Added in the Presence of Multinational Firms with an Application to High-Tech	OECD Analytical AMNE database	Current TiVA analysis misses about 10 percent of FDI-related GVC activity in terms of global GDP; this paper provides a framework for accounting for FDI in GVCs.	Understanding the importance of FDI is important for bilateral trade treaties.

(Continued)

Appendix C Continued

Chapter	Data Sources	Measurement and Valuation Issues	Understanding GVCs and Policy Issues
4. Trade in Services, Intangible Capital, and the Profit-shifting Hypothesis	Italian firm-level data from the Bank of Italy	Highlights need for additional understanding of where intangible capital, such as an intellectual property product, is produced versus assigned.	Tax policy and whether firms are shifting profits to avoid taxes.
5. Talent, Geography, and Offshore R&D	Orbis data merged with PATSTAT data (37 countries)	Highlights the need to identify firms by both nationality and location.	Should countries subsidize domestic innovation or encourage foreign affiliates? Implications for restrictions on immigration and trade on labor productivity within a country.
6. The Nature and Direction of Innovation in Global Value Chains for Wind-Energy Technologies	Wind-energy reports from Navigant; wind-energy needs from IEA	Case study for wind-energy industry, which may help improve official statistics.	Does the location of suppliers affect the temporal component of innovation (in other words, does offshoring innovation decrease long-term innovation)?
7. Economies of Scope and Relational Contracts: Exploring Global Value Chains in the Automotive Industry	U.S. Customs microdata (LFTTD) (U.S. and Japanese automobile manufacturers)	Case study for automobile manufacturers using microdata; ownership identified by manufacturer ID to help understand buyer–supplier links.	Trade policies are partly determined by economies of scope among suppliers. Provides insight into the organizational strategy of firms in the United States versus Japan. These data may help understand the resiliency of GVCs in the wake of supply shocks, such as those resulting from the COVID-19 pandemic.

8. Foreign Direct Investments and Superstar Spillovers: Evidence from Firm-to-Firm Transactions	National Belgian Bank (NBB) business-to-business transaction dataset merged with Central Balance Sheet company accounts data; FDI info from NBB FDI survey; Intrastate trade survey; customs data	Evidence regarding value of superstar value chains.	Informs government policies intended to attract MNEs by providing information about spillovers to the host country from foreign MNEs.
9. Creation and Diffusion of Knowledge in the Global	PATSTAT and USPTO merged with firm and affiliate information from Orbis and Orbis IP databases	Merging IP data from patent offices with Orbis provides information on inventors' location, gender, and other characteristics.	Provides information about inventor mobility and ability of MNEs to diffuse knowledge across borders; time zone differences and physical distance are barriers to collaborative innovation.
10. Firm Selection and Organizational Choice: Complex Patterns of Global Sourcing	Survey data collected by Statistics Denmark (3 waves)	Manufacturing offshoring surveys provide critical information about type of activities that are offshored and changes in activity over time; provides information on inter-firm versus intra-firm sourcing.	Provides information on complex outsourcing patterns to help understand firms' make/buy decisions.

(Continued)

175

Appendix C Continued

Chapter	Data Sources	Measurement and Valuation Issues	Understanding GVCs and Policy Issues
11. Are Customs Records Consistent Across Countries?	U.S. customs import microdata (LFTTD) and Columbian export data	Understanding the accuracy of firm-to-firm transactions data in international trade and GVCs; there is a growing gap in aggregate trade data, especially for smaller value shipments and nonwholesale retailers.	Using firm-to-firm networks and GVCs could find imaginary links in networks and may be not be correctly measuring the duration of relationships. These data are used to enforce commercial policy, but poor data may make it difficult to identify bad actors.
12. Capital Flows in Global Value Chains	BEA Capital Flow Tables and World Input-Output Tables	Understanding capital flows in GVCs.	Gains from trade liberalization may not be captured using existing frameworks.
13. Colocation of Production and Innovation: Evidence from the United States	Longitudinal Business Database (LBD), NAICS classifications and establishment geocodes from the Business Register; LFTTD Database; R&D surveys, USPTO database.	Merging multiple data sources provides location data to help understand trade in IP within firms.	Relationship between production and innovation; can help understand how policy-induced changes to R&D, such as R&D tax credits, affects manufacturing decisions.

NOTES: AMNE = Activity of Multinational Enterprises; EPO = European Patent Office; FDI = foreign direct investment; GDP = gross domestic product; GVC = global value chain; IEA = International Energy Agency; IP = intellectual property; LFTTD = Longitudinal Foreign Trade Transactions Database; MNE = multinational enterprise; NAICS = North American Industry Classification System; OECD = Organisation for Economic Co-operation and Development; PATSTAT = Patent Statistical database; R&D = research and development; TiVA = trade in value added; USPTO = U.S. Patent and Trademark Office.